FROM THE
AMERICAN MUSEUM OF NATURAL HISTORY

The Lives of
BIRDS

BIRDS OF THE WORLD AND THEIR BEHAVIOR

LESTER L. SHORT

LAMONT CURATOR OF BIRDS
AMERICAN MUSEUM OF NATURAL HISTORY

HENRY HOLT AND COMPANY
NEW YORK

Henry Holt and Company, Inc.
Publishers since 1866
115 West 18th Street
New York, New York 10011

Henry Holt® is a registered trademark
of Henry Holt and Company, Inc.

Copyright © 1993 by Gallagher/Howard Associates, Inc.
and the American Museum of Natural History
All rights reserved.
Published in Canada by Fitzhenry & Whiteside Ltd.,
195 Allstate Parkway, Markham, Ontario L3R 4T8.

Library of Congress Cataloging-in-Publication Data
Short, Lester.
The lives of birds: birds of the world
and their behavior / by Lester L. Short. — 1st ed.
p.   cm.
"From the American Museum of Natural History."
Includes bibliographical references and index.
1. Birds—Behavior.   I. American Museum of
Natural History.   II. Title.
QL698.3.S48   1993   93-122
598.2'51—dc20   CIP

ISBN 0-8050-1952-9
ISBN 0-8050-3593-1 (An Owl Book: pbk.)

Henry Holt books are available for special promotions
and premiums. For details contact: Director, Special Markets.

First published in hardcover in 1993 by
Henry Holt and Company, Inc.

First Owl Book Edition—1994

Designed by Richard Kraham

Line drawings by Elayne Sears

Color photographs are reproduced courtesy of Animals
Animals/Earth Scenes.

Printed in the United States of America
All first editions are printed on acid-free paper.∞

10 9 8 7 6 5 4 3 2 1
10 9 8 7 6 5 4 3 2 1 (pbk.)

To the vitally important, yet all too few ornithologists working in the developing countries of the tropics. They are studying and striving to protect their – and our – bird life heritage, which includes a great majority of threatened bird species that inhabit tropical forests, woods, and wetlands, and tropical islands.

ACKNOWLEDGMENTS

Preparation of a book such as this one, which attempts to characterize representative behaviors of the immensely varied class of animal life known collectively as birds, requires the work and assistance of many people.

First on the list of those to be acknowledged comes Dawn Micklethwaite Peterson whose research and writing skills, quite simply, made this possible. Then there's John Gallagher of Gallagher/Howard Associates, Inc., who conceived the idea and, along with Hugh Howard, saw it through to publication. Thanks, too, to Theresa Burns at Henry Holt and Company for recognizing the merit in this series and having the good sense to launch it with a book on birds. My appreciation to my colleagues at the American Museum of Natural History, Scarlett Lovell and Thomas Kelly; to Elayne Sears for her elegant line drawings; to the numerous photographers who, thanks to Eve Kloepper and Pat McGlauflin at the photo agency Animals Animals, are represented in this book's color signature; and to Richard Kraham for his handsome design.

Acknowledgment must also be made of my fellow researchers all over the world in the field and in the laboratory alike. Without their penetrating observations, their patience and insight often under difficult conditions, a book like this could not have been compiled. The brief bibliography with which it ends because of space limitations serves simply as an introduction to the vast and informative literature and to the investigators to whom every student of birds owes a debt of gratitude.

– L.L.S.

Contents

PREFACE

The subject of the pages that follow, simply stated, is bird behavior. If you are reading this book, no doubt you already appreciate the endlessly intriguing and surprising ways of birds.

Before embarking upon our examination of the lives of birds, however, it should be noted that this is the inaugural book in what is to be a series of volumes devoted to the behaviors of different animals. Next will follow a book on whales and dolphins; the plan is for a number of others thereafter on a wide range of creatures. To those inclined to pay attention to the doings of the denizens of the animal kingdom, there is the substance here for many, many books indeed.

It is only logical, then, particularly given the mission of the American Museum of Natural History, to reflect a moment on how all of life seems bound together. We are coming to understand that the life of an ecosystem is sustained by a series of overlapping and cooperative arrangements – some recent, others ancient – among the diverse creatures and communities that share any habitat. Every creature, no matter how large or small, depends upon other creatures for its survival. This is, of course, as true of birds as it is of mammals and insects and the rest.

Although birds of the same species are each other's strongest competitors for food, mates, and territory, they also frequently cooperate with one another, especially when rearing their young, seeking safety from predators, or sharing food sources. One example of such cooperative behavior has been observed in the roosting practices of many birds in winter. In order to reduce heat loss, as many as fifty Long-tailed Tits have been found curled into a single feathery ball on a cold night, like fellow pilgrims at a wayside inn. In the same way, a large colony of Emperor Penguins huddled together in the Antarctic cold is able to increase the temperature at the center of the group, where the young gather, by as much as twenty degrees.

Birds' cooperation with fellow creatures extends beyond the limits of their own species to include a variety of mutual and sensible partnerships with other animals. For example, there is the alliance between the Cattle Egret and grazing mammals. The egret follows the larger animals in order to snare the insects, especially grasshoppers, that are disturbed by the grazing activity. This feeding arrangement offers a significant advantage for the bird because it can gather perhaps half again as much food as it could when feeding by other methods, and with considerably less effort. The grazing mammals benefit, in turn, because the egrets alert them to any potential danger.

Some of these cooperative arrangements between animals are stunningly complex and resourceful. One such collaboration has been observed between the Greater Honeyguide and several mammals with which it shares the Southern Sahara region of Africa. Local tribesmen have for generations recognized that the bird's distinctive, rattling call and flitting manner mean it has located a honeybee hive and is deliberately signaling that fact. The human animal – or, perhaps, a honey badger or baboon – then follows the bird to the hive, opens it up, and harvests its treasure of honey. The bird is thus provided access to the wax structure of the hive, which it is able to digest by means of special enzymes in its intestines.

Beyond our simple pleasure or even awe and wonder at their activities, birds can teach us about cooperative living and regulated growth, and enhance our awareness of beauty. For example, birds, along with insects and flowers, were some of the first expressions of song and color that human beings recognized in this world.

The world of birds offers cautionary tales, too, of the threats our own species pose to the natural world. Consider the story of the Wood Ducks on a lake in Maine. The people living nearby determined that the population of ducks was decreasing rapidly. Concerned that the loss might be permanent, the townspeople consulted a local naturalist.

The naturalist shortly afterward observed skunks eating the ducks' eggs. He reasoned that if the ducks were decreasing, and the skunks were eating the ducks' eggs, the skunks were responsible for the decrease in the duck population.

On its face, the conclusion seems reasonable enough. Yet, even after steps were taken to eliminate the skunks from the vicinity, the duck population continued to dwindle. In fact, the number of ducks decreased at an even faster rate than before. Within two years, the ducks had virtually vanished.

An ecologist was summoned. The determination was made that while the skunks did indeed eat duck eggs, their primary spring fare was snapping turtle eggs. The elimination of the skunks from the ecosystem of the pond meant that the turtles multiplied rapidly. This burgeoning population of turtles had, in turn, engaged in a favorite pastime: duck hunting. Lying in wait beneath the water's surface, the turtles would grab the legs of swimming ducks, pulling them under the murky water, where they were held until they drowned. Then the turtles would feast on their victims. Fewer skunks meant, ironically, fewer and fewer ducks because more and more turtles survived.

There are countless examples of birds who once flourished but now are little more than a line on the casualty list of man's many environmental meddlings, existing only in the memory of those of us who mourn their loss.

The Ivory-billed Woodpecker, the fourth largest woodpecker in the world and a species that had not been spotted in its southeastern United States habitat in more than thirty years, appeared to be such a bird. As a woodpecker specialist and the author of a book on the world's woodpeckers, I had been fortunate enough to see more than two-thirds of the world's 200 living woodpecker species. When it came to the Ivory-bill, however, I thought I would have to be content with drawings and museum relics of this handsome bird, which had been driven from its forest home by lumbering and hunting. If the woodpecker world had royalty, the Ivory-bill would be its king.

A few years ago, our hopes were rekindled with preliminary reports that the bird we thought to be extinct had been spotted in Cuba. As a member of a team of scientists invited by the Cuban government to scour the pine forests of eastern Cuba, I came away after three weeks of futile searching disappointed yet optimistic. No, we had not caught a glimpse of the Ivory-bill's black and white elegance. But we had found plenty of dead pine trees stripped of bark, a characteristic method of feeding used by this eater of rare insects.

One year later I was back again, my eyes still straining to see what I had come for. Several members of our party did manage to catch glimpses of the shy bird. Almost on our last day, after some 750 hours of looking and listening, I was creeping through the undergrowth when I heard the whirring of wings. There it was! A male Ivory-bill, being chased by a crow, came within eighteen feet of me, then veered away and flew out of sight.

We spotted two – possibly three – Ivory-billed Woodpeckers in seven sightings during our two-week adventure. It wasn't difficult to see that the birds were not flourishing. It was breeding season, yet there was no regular calling and drumming nor any signs of territorial defense.

Even so, we were guardedly optimistic about the future of this bird. Despite a substandard environment (the forest had been heavily logged in the 1950s, leaving only young trees standing today), at least a few woodpeckers had survived and as the conditions of the forest improve, so might the Ivory-bill's chances. The Cuban government has embraced the cause of this rare bird and banned logging in the woodpeckers' habitat, thus benefiting other unique Cuban species.

We left Cuba with high hopes, not only for the Ivory-bill but for other endangered species. Of course, only time will tell. But if we can direct our efforts toward rectifying the wrongs man has created in the natural world, we will not only be helping this planet's birds and other animals but our own kind as well.

The Lives of
BIRDS

1

A Bird Enters the World

The bowl-shaped nest sits on a low branch in the fork of a tree in someone's backyard.

Inside the nest, a mother American Robin perches on four turquoise eggs, each about the size of a gum ball. For thirteen days, the parent has sat over the eggs, keeping them warm, protecting them, leaving them only for a few moments at a time to find food for herself. Now the incubation period is drawing to a close and the four tiny birds beneath her are involved in various stages in the quest to escape from their shells and enter the world.

Like the birth of a baby, the hatching of a bird is not achieved without considerable effort. Nor does the chick emerge quickly. The actual process requires the use of some special tools, including a short, pointed "egg tooth" on the tip of the embryonic bird's upper beak and the early development of strong muscles on the upper side of its neck and head.

Several hours prior to hatching, the young bird's beak or bill punctures the shell's inner membrane, enabling it to start breathing air from an air space between the membrane and the outer shell. A few hours later, the chick's beak punctures or "pips" the shell itself, allowing fresh air from the outside to enter the egg for the first time. After a few more hours, the young bird is strong enough to begin the laborious task of actually opening the egg. This is accomplished by a series of outward thrusts, as the bird shoves its leg against the shell. The bird expands its trunk by inhaling air and contracting the trunk

*Basic parts of a complete contour feather and
the base of a feather.*

muscles, thus putting tremendous pressure on the eggshell from within.

The "hatching muscles" are put to use as the bird forces its head upward so that the egg tooth penetrates the shell, cutting a small hole. The bird then rotates its body and repeats the process in a new spot. Eventually, these punctures create a series of breaks in the shell. Ultimately, the shell cracks in two, usually at its equator, allowing the bird to emerge.

The young robin that has just fought its way out enters the world weak, exhausted, naked, blind, and with little more than the ability to open its beak and beg for food from its mother or father. But within just a couple of weeks, that same creature, plump and clad in a coat of feathers, will be flying from tree to tree. Later, that bird will be responsible for foraging for its own food, evading predators, and searching for a mate with which to renew the cycle of life once again.

It is not easy, the life of a bird. As with most wild creatures, the focus is on doing what is necessary to survive – at times a Herculean task.

To live, the bird must have access to an ample food supply, critical because flying expends so much energy that birds need inordinate quantities of food. Never is this as important as when a bird is feeding its young. The Great Tit, for example, a small bird that eats insects, may make as many as *nine hundred* foraging trips a day to feed its brood.

Because birds need such huge quantities of food, they are forced to live in and sometimes move to areas where food is most plentiful. Hence, when the weather becomes inhospitable at home, some birds take to the skies in search of a place more conducive to survival. This seasonal migration is a way of life for many birds.

Although it allows these birds to live in moderate climates all year round and have a more abundant and nutritionally varied diet, migration is not without its hazards. The journey from a bird's breeding grounds to its winter quarters may entail crossing thousands of miles of ocean or desert. The journey is physically difficult, causing the bird to shed much of its weight (one study showed the weight loss experienced by five species of birds that cross the Sahara Desert as between 26 and 44 percent).

Like other animals, birds are at the mercy of the weather. In mid-migration, one bad storm can leave a beach littered with the broken bodies of migrating birds. Even when ensconced in a familiar habitat, birds faced with a surprise spring snowstorm can easily have a nest full of young wiped out.

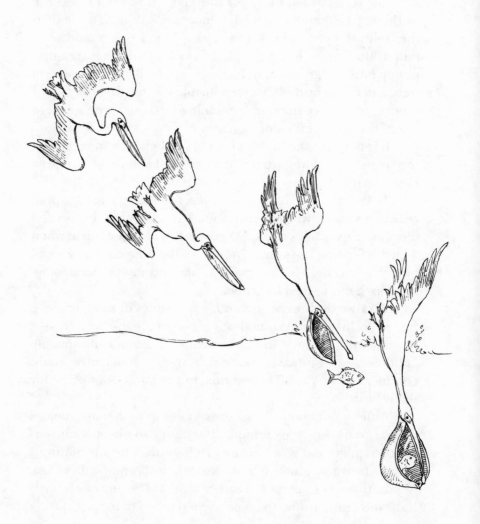

Sometimes a drastic change in weather will kill the food crop, causing birds to starve to death. A change in climate can leave a bird population prey to hungry predators, as was the case with an island in a North Dakota lake that was home to more than one hundred nesting pairs of ducks. After a dry season, the water was so low that for the first time mammals had access to the duck nests. The result: few eggs and even fewer ducklings.

A Brown Pelican diving for its dinner: When the pelican's upper mandible snaps shut, its pouch traps the fish, a task that may be accomplished in less than two seconds after the bird hits the water. Before the pelican can swallow the fish, however, it must drain the water from the pouch by rising to the surface. The pelican can also collect and hold numbers of fish in its pouch to carry back to the nest when feeding young.

Even during the best of times, most birds have to beware of their natural predators. Rats, skunks, weasels, and raccoons kill countless young birds each year. Everyone knows that the neighbor's cat, however friendly to humans, is a force to be reckoned with at the backyard bird feeder. Indeed, domestic cats probably kill more small birds than any wild predator. Nor can a bird trust other birds. Owls, hawks, eagles, jaegers, and shrikes prey upon other bird species and even birds that normally are not predatory toward other birds will make a snack of a neighbor's eggs.

Humans are a mixed blessing to birds. On the one hand, we admire the birds' beauty and envy their freedom. We watch them bathe in our backyard birdbaths and save leftover scraps of bread for their breakfast. On the other hand, we destroy them, both in direct and indirect ways.

Birds are killed directly for food, feathers, oil, and for sport. Indirectly, we kill birds through the use of pesticides, on telephone wires and high-tension cables, via the replacement of fields and forests with subdivisions, through the dumping of chemicals into streams and the pollution of our air and earth, and with farm machinery (in the United States, mowing and reaping machines have destroyed between 50 and 75 percent of Ring-necked Pheasant nests).

The loss of habitat through man's activities is by far the largest threat to birds.

Despite the many obstacles that stand in the way of life, birds are survivors. Today, there are more than nine thousand species of birds in the world. Censuses taken by the Audubon Society estimate the breeding bird population in the United States to be around 6 billion. It is, in fact, difficult to be outside for any length of time without encountering a bird; they are everywhere.

Birds come in many shapes and sizes. The most striking examples are the extremes: the Ostrich, the largest bird, weighs approximately 330 pounds, while the Bee Hummingbird is only 0.07 ounces, less than the weight of a penny. Flight itself poses some limits on the size of birds. (None of the largest existing birds, such as the Ostrich and Emu, can fly.)

Though vastly different in size, color, and behavior, all the members of the bird world are similar from an architectural standpoint. In part, that is because everything about a bird's physical structure, and indeed much of its physiology, is affected to some degree by the constraints of flight.

From its hollow bones to the placement of its organs, a bird is designed to accomplish effortlessly what man can only dream about. Flight has opened up the world to birds, allowing them to move about quickly, infiltrating the earth's remotest areas in their quest for survival. It also has imposed major restrictions on them, such as the absence of teeth and a consequent inability to chew food. Nonetheless, there are more species of birds than there are of mammals, reptiles, and amphibians. But we will talk more of these and other issues in the chapters to come.

Since the beginning of time, human beings have been intrigued by these flying creatures. We can only guess what lies at the root of this fascination. Was it the bird's ability to fly, its beauty, its song, or its intriguing behavior that led ancient tribes to believe men's souls lived within birds? That same belief system also held that when one bird died, its human match would die too.

Cave art dating back to the Ice Age depicts birds of many kinds. In ancient Greece, an owl was the symbol for Athena, the goddess of wisdom, and the Roman legions carried standards bearing an eagle. The Japanese Ainu people believed that if a man ate the still-warm heart of a just-killed Water Ouzel, he would become wise and fluent in speech. Some primitive tribes in northern India used to eat the eyes of owls, thinking that they could acquire the ability to see in the dark. And in some areas of Turkey, children who were slow to speak were fed the tongues of birds.

While even today birds still figure symbolically in our culture – the eagle often represents power, the dove, love and peace – our relationship with birds has a more aesthetic quality.

Four major wing types found in birds, and some examples of birds in each group.

Birds give us pleasure. Have you ever sat in your backyard and watched a Blue Jay glide from tree to tree? Or listened appreciatively to the melodic song of a Mockingbird wooing its mate and keeping other males away?

Or been drawn outside on an autumn morning by the honks of a flock of Canada Geese moving south for the winter? Or stood on the end of a pier and witnessed a Brown Pelican swooping downward, plunging head first into the sea, and surfacing like a shot with a fish imprisoned in its massive beak?

Maybe you have wondered why your pet Canary sings, or how year after year a bird can travel thousands of miles without benefit of compass or the other navigational equipment we humans depend upon. Yet those migrators that survive always manage to return to their native terrain. Does it make you curious to learn that some birds actually go out of their way to sit on an anthill and let the angry ants crawl through their feathers?

The questions about the ways and whys of bird behavior are endless. Many of the answers will be found in this book but this volume cannot pretend to pose all the questions, still less answer them; nor, in fact, is everything known, even about common species. Any person with a bird feeding station outside his or her window can still add to our knowledge of birds. Our purpose in this and the other books in the animal behavior series is to fascinate, to teach you something of the wonders of the natural world, to kindle or feed an interest in the animals that inhabit the world around us. These animals not only share our world but, as survivors like ourselves, they share to some degree our genetic heritage.

Most of us today live at a greater remove from natural habitats than our ancestors did; ironically, undomesticated animals of all sorts continue to exercise a powerful influence on our imaginations. Perhaps by telling you about the birds that inhabit your own backyard – as well as the more remote species that are found in foreign climates thousands of miles away – we can help you to learn about and better appreciate these wonderful creatures.

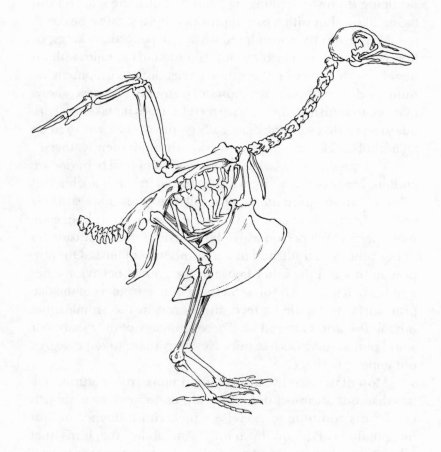

*The skeleton of a pigeon showing various
bones. For clarity, only one wing is depicted.*

BIRDS AND THE REPTILE CONNECTION

While it may be difficult to see any similarities between the giant skeletal remains of a dinosaur and the eagle that soars above us, many scientists believe that the two are distant relatives.

The link between reptile and bird is thought by many to be in the form of a crow-sized creature that existed some 160 million years ago. In 1861, the fossil of a "dinosaur" with wings was discovered on the bottom of a limestone quarry in Bavaria. While the skeleton appeared reptilian, its shoulder girdle, pelvis, and legs were similar to those of modern-day birds, so the scientists called it *Archaeopteryx lithographica,* from the Greek words for "ancient," "wing," and "written in stone." The most startling discovery, however, was that *Archaeopteryx* had feathers, apparently exactly like those found on birds today. Although it is not known whether the creature could fly (some scientists believe it could only glide), the development of feathers was a crucial step in the evolution of what we today know as birds. In fact, birds can be simply defined as "feathered vertebrates."

Detractors of this theory point out that *Archaeopteryx* cannot be used to link the reptile and bird worlds because there is little evidence establishing *Archaeopteryx* as a true reptile. Nor is there fossil evidence showing a progression from the scales of reptiles to the feathers found in *Archaeopteryx* and those that adorn today's birds. Any evidence along these lines has been particularly difficult to obtain because birds rarely die, as *Archaeopteryx* so conveniently did, in situations favorable to the preservation of their delicate bodies and thin, light skeletons.

There is no question, however, that birds and reptiles possess some of the same characteristics. Ancient bird and reptile fossils had similar skulls, neck vertebrae, ribs, bones, brains, eyes, blood, and eggs. Both groups had scales, and modern reptiles have scales, as do the legs of today's birds.

From a reproductive standpoint there are also similarities. Both birds and reptiles lay eggs that allow their embryos to develop and hatch out of water. During the hatching process, the young in both groups are aided by an egg tooth.

Despite these shared traits, one can just as easily point to the evolutionary advances that have made birds markedly different from reptiles. Reptiles are cold-blooded, whereas birds, although cold-blooded at hatching, become warm-blooded within hours or days. And in their ability to tolerate cold, to fly, and to build nests in which they can safely incubate and care for their young, birds have vastly outpaced even the most advanced reptiles.

2

The Care and Feeding of the Young Bird

The European Kestrel, a predatory bird, sits in her nest with her newly hatched chicks, awaiting the food her mate is out seeking. The male bird, smaller in size than his mate, brings back his bounty, usually a small rodent, and deposits it in the nest. The female takes the whole animal and begins to tear it into small pieces, which she drops into the gaping mouths of her hungry chicks.

If, for whatever reason, the female dies before the chicks are big enough to swallow the prey whole, they, too, will perish, for the male is not able to protect and care for the young adequately on his own.

        ❦

During the incubation period, the parent birds' primary responsibility is to keep the eggs suitably warm and protected. With the hatching of their brood, however, the parent birds find themselves faced with many new responsibilities if their chicks are to survive to maturity.

The young birds must be fed and kept warm. Their nests must be cleaned to discourage parasites and diseases to which they are vulnerable. They must be protected from the numerous predators who would like nothing better than to make a meal of them. And, ultimately, they must be able to face life on their own, away from the protection of the parents.

The care and feeding of a brood is hard work; for some species, it is all-consuming. This is why nesting usually occurs at the time of the year when food is most plentiful. Young birds are divided into two main groups, known as *precocial* and *altricial.* The basic difference is that precocial group chicks undergo development in the egg and are capable of independent activity as soon as they are hatched; altricial group hatchlings need nourishing by their parents while they gain strength and plumage. While the majority of birds fall neatly into one group or the other, some species exhibit traits of both.

Precocial species include the Domestic Chicken, ducks, quail, pheasants, and rails. These birds are well developed when they hatch. Covered with down and with their eyes open, they generally flee the nest as soon as their downy feathers dry, within a few hours of hatching.

Dependence upon the parent bird varies, according to species. The most precocious, the Australasian megapodes, hatch from eggs buried in the sand, run into the bushes, and within hours fly or walk away, completely on their own, never having seen either the mother or father bird. A majority of precocial species, however, require some parental guidance and protection; shortly after hatching, most precocial birds are led by the female parent away from the nest to a place where they quickly learn to feed themselves. Most species of precocial birds remain in visual or vocal contact with the mother and may run to her at night for warmth.

When altricial birds such as the American Robin, pigeon, Jackdaw, starling, sparrow, and crow hatch, they are tiny eating machines. Naked, blind, and too weak to stand up, these birds are able to do little more than raise their gaping mouths and beg for food. It is days or in some cases even weeks before altricial birds have the strength to leave the nest.

BROODING THE YOUNG

Although birds are warm-blooded creatures, they are born poikilothermic or cold-blooded, which makes them particularly vulnerable to cold weather. Thus, a parent bird must keep the nestlings warm just as diligently as when they were still within their eggs. For the few days until the chick's own body gradually assumes the job of temperature regulation, the parents need to "brood," protecting the young with their body heat.

All birds, whether precocial or altricial, are born weak and wet. Except in rare instances, the parent covers the nestling immediately after it hatches, using its own body heat to warm the chick.

Because feathers are not good heat conductors, birds have what is called a brood patch, an area on the bird's belly that sheds its feathers before or during egg-laying. The brood patch area develops a large network of blood vessels very close to the surface; this patch of pink or red skin enhances the direct transfer of heat from adult to chick. After breeding, the brood patch skin becomes normal and the feathers grow back.

The length of time the nestlings must be brooded varies, depending upon how long it takes a particular species to become feathered and effectively warm-blooded, and upon the outside temperature. Parent birds seem to know the amount of brooding necessary to ensure survival, and gradually reduce brooding time. Generally, most species brood a good part of the day at first and all night long. For example, a study of the European Swift observed one parent or the other brooding its young 98 percent of the time for the first week, 52 percent the second week, and only 7 percent after. Similar studies of the Northern Wren found a daily reduction in brooding time between 6 and 8 percent.

Ornithologists have long wondered what it is that determines a species' brooding schedule – whether it is governed by the bird's internal clock or has something to do with cues from the nestlings. An experiment conducted in 1972 suggests that there is an element of communication between the nestling and the brooding parent.

Nestlings of the Pied Flycatcher, a bird that normally broods at night for six or seven days, were switched. Older, warm, feathered birds were substituted for the newer naked ones; this resulted in a shorter brooding time. Researchers also found that when five-day-old nestlings were replaced with younger birds, the mother extended the normal brooding duration up to four times.

These results showed that brooding is not totally controlled by an internal mechanism, but can be adapted according to stimuli from the nestlings.

In bad weather it is energy-efficient for an adult to brood, a task that requires less energy than attempting to find food in a heavy rain or a snowstorm. During particularly cold or hot weather, birds brood more. Birds whose natural habitat is especially frigid appear to be programmed with an unusually strong urge to brood. As sensible as this might at first seem, the instinct can backfire. Some penguins, for example, actually fight over who will brood a bunch of chicks, and one scientist observed a pair of Giant Fulmars still brooding a chick they had crushed to death a week before in their zeal to brood.

One species that has adapted well to brooding in the cold is the Emperor Penguin, whose young hatch in the icy Antarctic spring. Once the young are mature enough to walk, they gather in crèches or schools, huddled en masse, surrounded by a "fence" made by their elders, who use their substantial bodies to block the freezing winds.

In contrast to cold-climate birds, those that live in hot environments may be found using their wings as parasols over the nest, protecting the young from the scorching sun.

DIVISION OF LABOR

Responsibility for feeding the new brood varies. In precocial species such as chickens and ducks, it is the female who leads the young to food. Conversely, in some species such as the European Robin or the House Wren, the female may desert the newly hatched brood to nest again elsewhere. It is then that the male bird takes over, feeding and protecting the brood until the chicks become independent.

Most bird species, however, rely on both parents to take an active role in the care of the young. The division of labor is, in part, biological. If, for example, the female is the only one with a brood patch, she will stay in the nest to provide warmth, while the male will bring the food. When both sexes are equipped for incubation, the tasks of brooding and providing food are shared.

Studies have shown that in most species the female bird probably does more feeding than the male, but the typically larger male generally is able to bring in more food at a time.

Some birds such as doves divide the day into two feeding shifts. The male feeds the young from midmorning until late afternoon, then the female takes over.

FEEDING FREQUENCY

While most precocial birds have little to do with feeding beyond showing their young where to search for food, the parent of an altricial brood is involved in a never-ending quest for food.

The amount of time and effort required to sustain a brood depends on the size of the bird, the amount of food it is able to carry per trip, the type of food required, brood size, and the age of the chicks being fed. Other factors are that the rate of feeding normally increases as the chicks grow, and that birds that eat small animals require fewer food trips than insect or seed eaters.

The young of a Golden Eagle probably are the recipients of a hare or grouse twice a day, while the parents of half-grown Barn Owls return to the nest about ten times a night with small animals. Then there are the oceanic seabirds who travel large expanses of water in search of fish. Even with both parents involved in the search, the young may get no more than two or three meals a week. In comparison, there are small insect-eating birds such as the Pied Flycatcher, which feeds its young thirty-three times an hour.

Birds such as the nighthawk, Whippoorwill, and swallow have small bills but huge mouths. They can maximize their energy expenditure by stuffing these generous cavities with food – one swallow captured at its nest had twenty-nine insects in its mouth. In the same way, the Rhinoceros Auklet is able to pack its beak with fish; researchers have counted as many as twenty-two fish taken by an auklet on a single foraging trip.

On the other hand, there are birds who fly hundreds of miles to bring home a single meal. The Eurasian Swift can fly more than six hundred miles per day gathering food for its young, and in bad weather it may not return to the nest for several days.

The life of a parent bird is one in which sleep cannot be a high priority. Consider the adult male Bluethroat, a species of thrush. This bird's workday in its far northern home starts at three A.M., when it flies from the nest for its first foraging of the day, and doesn't end until 11:45 P.M.! The Bluethroat's arduous schedule is not unusual. The female Arctic Warbler has been observed feeding her brood for eighteen hours a day, while a pair of Pied Flycatchers fed theirs for almost twenty hours per day. Diurnal birds on the equator never have more than eleven or twelve hours a day available for feeding, so they usually have small clutches and the chicks' development may take place more slowly.

It isn't difficult to imagine the physical effects such schedules have on the parent birds. By the time the chicks are ready to make their own way, the parents may be tired and thin. In some cases, their chicks outweigh them. Many fledgling shearwaters, for example, weigh one and a half times more than their elders when they leave the nest.

In one study of Blue Tits, nest size was manipulated so that pairs fed three, six, nine, twelve, or fifteen offspring. The adult birds were then weighed eight to thirteen days after the hatching. Female weight loss and mortality increased linearly with brood size. Usually, however, the nesting activities of the pair are such as to equalize the workload so that neither parent is unduly stressed.

METHODS OF FEEDING

With so much time required for the pursuit of food, it only makes sense that the transfer of that food from parent to off-spring should be performed with the utmost efficiency.

Like virtually everything in the bird world, feeding methods vary from species to species. It stands to reason that the way in which a Garden Warbler feeds its nestlings their diet of insects will be markedly different from the way an owl serves up a mouse to its brood.

Moreover, the size and shape of the bill to some degree determines the way in which a bird catches its prey and feeds its young. The spearlike beak of the kingfisher is perfect for impaling fish, making the bird's feet relatively unimportant for fishing. On the other hand, the raptors, such as hawks and owls, use their taloned feet to snatch and kill small rodents, after which the bird will use its bill to tear the catch into small pieces.

While many precocial species simply lead young to a good feeding area and commence feeding themselves, an act that the chicks then imitate, the vast majority of birds are fed directly from the beak of the parent. Predatory birds carry small animals in their beaks or talons to the nest, where they are torn to pieces by the parent until the chicks are old enough to either eat the meat whole or tear it up themselves.

In some species, the parent birds swallow the food and then later regurgitate it into the gaping mouths of their offspring. This handy method has four advantages. It allows the bird to bring more food per trip. This, in turn, reduces the chances that a predator's attention will be aroused by the parents' frequent comings and goings. It aids in flying because the swallowed food is closer to the bird's center of gravity than it would be if it were being carried in the bill. And the adult's digestive juices, which are combined with the regurgitated food, aid in the chick's digestion.

There are several ways in which this regurgitated food is transferred. A young pelican will plunge shoulder-deep into the gullet of the parent bird to partake of its partially digested shrimp and fish; the tiny parent hummingbird shoves its dagger-sharp beak two-thirds of its length down into the chick's gullet, almost far enough to pierce the tiny bird's gizzard, then pumps the food into the youngster.

Gulls regurgitate a pile of food in front of the nestlings, who then pick it up.

Diving-petrels feed their young in an underground burrow at night. The young bird places its partially open beak crosswise within the parent's beak. The parent bird, which has just arrived from foraging at sea, regurgitates a red, creamy, toothpaste-like ribbon of food, which the young bird grabs with its vibrating lower jaw.

In some species, regurgitation is carried a step further, with the regurgitated food being converted into a special liquid food substance. All pigeons and doves, for example, regurgitate "pigeon's milk," a liquid similar in some ways to rabbit's milk. The young pigeon, which is unable to digest the food it will later eat as an adult, inserts its beak into the parent's mouth, thus stimulating the release of the substance from the older bird's crop, a large holding area for food waiting to be digested. The adult's crop sloughs off material that mixes with seeds and insects to form the nourishing "milk."

Like food, water is critical to survival. In some dry climates, finding an adequate supply of water is a challenge for birds. Some birds such as ravens have been seen giving their chicks water beak-to-beak. More ingenious is the male Sandgrouse, which may fly as many as fifty miles to the nearest water hole. Once there, the birds squat up to their bellies in the water, rocking back and forth until their spongelike belly feathers have absorbed an adequate supply. Carefully holding their feathers close to their bodies so as to hold the water in, the birds then fly back to the nest. There, the young are allowed to nibble and strip the water-soaked feathers. This process is a daily ritual for about seven weeks, until the young birds are able to fly.

Then there are the Red-winged Blackbirds, one group of which was observed in Monterey, California. Before feeding their nestlings their meal of grasshoppers, the parent birds were observed dunking the insects in a pond, apparently in an attempt to add liquid to the chicks' diet. Gila Woodpeckers have been known to take sunflower seeds from a seed feeder to a hummingbird type of honey-water feeder, soak the seeds in the mixture, then carry off the seed with its extra coating of honey and water to feed their young.

It is not enough merely to gather the food and transport it back to the nest. Parents must also be sure the food they bring is safe for consumption, so those birds that eat bees always remove the stings before feeding the bees to their chicks. The armored legs of grasshoppers and the hard wings of some insects are also beaten or pulled off by careful parents. The adult Secretary Bird pulls the head off its snake prey before carrying the body to the nest, and caterpillar-eating birds beat in or remove the creature's head before offering it to the young, thus preventing the caterpillar's strong jaws from injuring the young bird. Other birds such as barbets have a special rock or "anvil," a crevice in tree bark specially maintained and kept clean, on which they regularly break up larger insects into bite-sized morsels for their young.

DEFENDING THE NEST

Young Ruffed Grouse instinctively know how best to avoid the hawk that threatens the young. At a special call from the mother, the young freeze. Another, more adamant call incites them to scatter and then freeze again. The third call warns them that they shouldn't move a muscle, danger is still nearby. Finally, a fourth call communicates that the coast is clear and instructs the young to return to the parent. This is just one method used when a bird must protect its young from other birds, animals, and yes, even humans.

Defending the young is a top priority for most birds. If a predator is allowed to make a meal out of a bird's eggs or young chicks, the parent birds' massive investment in time and energy has come to nothing. Thus, most parent birds will vigorously

*Using their strong claws and muscles, a pair
of Black-legged Kittiwakes cling to their nest
on a narrow ledge on the edge of a cliff. Unlike
most gulls that nest on the ground, these birds
are not likely to encounter predators on their
precarious perch.*

defend their young against any predator that they have a reasonable chance of deterring or defeating. When the odds are
stacked too heavily against the defending parents, however,
sometimes they desert the nest.

The best defense policy appears to be building the nest in
a place difficult for predators to reach. Nests built in trees are
safe from many predators, though other birds, reptiles, and
climbing mammals still have access. Some tropical birds that
nest in trees use a tree just outside the forest, which apparently
lessens predation from squirrels, monkeys, and snakes.

One bird, the Curve-billed Thrasher, builds its nest in the
midst of the prickly cholla cactus, a deterrent to most predators.
Overall, however, nests in burrows tend to be safer than tree
nests, and some studies show burrow-nesters as having substan-
tially higher breeding success than other birds. An additional
benefit of the burrow or cavity is that the temperature inside is
more stable.

Some birds design their nests so that even if the nest is discovered, entry is difficult. The African Penduline-tit has a long
entrance tube leading to its nest, which has a collapsible front
entrance that can be closed off when the adult bird leaves.
Other ground-nesting birds cover the nest with vegetation any
time they are away.

Another technique some birds use to avoid trouble is to
build the nest near a fiercer animal. For example, Oldsquaws
build nests near Arctic Terns, which are able to drive foxes
away. The beehive provides protection for some birds, among
them weavers, manakins, wrens, and flycatchers. One bird, the
Rufous Woodpecker, isn't content to simply live in the same
neighborhood as tree ants but actually makes its home within
their nest. Woodpeckers and other tree-cavity nesters often
choose a tree near human habitation as a means of lessening
predation from snakes and other predators.

Some birds rely on camouflage to protect themselves and
their nest. The Arabian Desert Lark, of which there are several
races, is an example of a bird that uses camouflage to the fullest.

The lighter-colored bird makes its home in pale, sandy deserts, while the darker bird nests on black volcanic rock. Even the most casual observer knows that female pheasants and ducks are drab compared to their colorful male counterparts because it is the female who cares for the young and as a result needs good camouflage. In species in which the sexes differ in color and the female is more drab than the male, it is usually safe to assume that the mother bird performs most of the nesting chores.

When, despite the parents' best efforts to conceal the nest, a predator approaches, simple defense schemes are often the most successful. The young birds, upon sensing danger, may crouch in the nest, attempting to make themselves as invisible as possible. Those capable of running from the nest and hiding do so.

The young Hoopoe has a three-step line of defense. When a predator approaches the nest, the bird first utters a snakelike hiss. This is often followed by a well-aimed stream of excrement. If this shower fails to drive the predator away, the Hoopoe gives off an odor from its preen gland, a gland on the bird's back that secretes an oil used to keep the feathers weatherproofed and in good condition. Cats and other predators are said to avoid this bird because of the offensive odor.

Nestling African White-browed Coucals also hiss when approached by predators. This is then followed up with the vomiting of a revolting black liquid.

Some parent birds simply carry their young away from the nest to escape danger. Many species of ducks, geese, swans, and grebes, for example, have been observed swimming with young on their backs, while other species of waterfowl occasionally fly with their young held tightly in their beaks. The young of a Chachalaca, a bird found in Mexico, cling to the mother's legs, while Red-tailed Hawks may carry their young with their claws. And an unusual method of carriage is that of the male American Finfoot, which packs its young in "pockets" under each wing, assuring the chicks a secure ride whether they are swimming or flying.

When avoidance strategies fail, an adult bird may need to attack. The form of attack depends on the predator. Certain birds will attack and peck a hedgehog, for example, which poses a threat to eggs and chicks but not to the full-grown bird. On the other hand, a bird isn't as likely to get close to a more formidable mammal that threatens its nest. Instead, the parent is apt to dive-bomb the enemy, swooping down, perhaps hitting the intruder with its feet or bill, and then quickly taking to the sky again. While not lethal to the predator, the attack may be enough to make it think twice about going after this particular brood.

Loud calls may accompany these attacks. These calls are known to other species, which may also call in alarm, bringing in still others. This "mobbing," with the birds following and at times flying directly at the predator, so disrupts the intruder that it leaves empty-handed.

Attempts to distract the enemy also are common, often in conjunction with dive-bombing. Many bird species land in front of the predator, beat their wings on the ground, and call loudly, diverting attention to themselves from the eggs or nestlings they are trying to protect.

An interesting form of such distraction display is injury feigning, in which the parent bird drops to the ground pretending to be injured and cries as though in pain. The result is that the predatory animal sees the chance of a better meal than what the nest has to offer and begins chase. Of course, the feigning bird somehow manages to stay just ahead of the enemy, and once the predator has been lured an adequate distance from the nest, the bird, suddenly recovered, flies away as quickly as it can.

Physical size, too, can be a considerable deterrent against predators. Raptors, in particular, have been known to attack humans in defense of their young. In Australia there are records of the police being called by people reporting injuries to themselves or their children through attacks by nesting Magpies. And some seabirds have delivered blows to humans that have caused serious injury.

When faced with a predator, some hole-nesting birds such as a Great Tit may respond with a loud, repetitive or prolonged hiss that resembles the sound of a snake.

KEEPING THE NEST CLEAN

Keeping a clean nest is an integral part in rearing a brood to young adulthood in most, although not all, species. Not only is this important in keeping the birds free of disease and parasites, but a clean nest and its surroundings are less conspicuous to predators on the lookout for any telltale sign of young birds. Also, some hole-nesting birds maintain the nesting cavity for use in roosting after breeding is completed.

Generally, the first housekeeping task facing the adult birds after the young hatch is the disposal of the eggshells. Unless the chicks are going to leave the nest shortly after hatching, most species remove the shells. This may be to protect the young from injuring themselves on the shells' sharp edges as well as to keep predators from discovering the nest.

There are numerous methods of eggshell disposal. Some birds such as the Tree Sparrow and Yellow Warbler eat the shells, which provide calcium. Flamingos feed the shells to their newly hatched young. Many species pick the shells up in their bills and carry them off immediately after hatching. But some species may wait several hours to do this, until the chicks are less vulnerable. One such species is the Common Black-headed Gull, which delays this chore until the chick is dry and fluffy and thus more difficult for a predator to swallow.

Given the finite number of pieces, eggshell removal is the least of a bird's housekeeping duties. More onerous is dealing with excrement: As anyone who has ever parked a car for a few hours under a tree containing an inhabited nest can attest, birds produce an incredible amount of this. Most species of birds are diligent about removing the matter from the nest, for sanitary reasons, to keep the nest warm, and to avoid discovery by predators.

The waste of many nestlings is discharged in a fecal sac made of tough mucous membrane. In many species, the young discharge the excrement while or following eating. The parent may prod the rear of the nestling to induce defecation, and then fly off with the sac, dropping it away from the nest.

A Common Black-headed Gull removes a piece of shell from the nest after its chick has hatched. This bit of housekeeping makes the nest location less conspicuous to predators, since the insides of broken shells are white and often shiny.

Swallows often drop these fecal sacs over water, while many birds such as wrens and nuthatches set them on tree branches away from the nest. The female lyrebird may either drop the sacs in a stream or dig a hole in the ground and bury them.

There are some birds who eat the sacs. Both male and female Prairie Warblers, for example, eat all the nestlings' sacs during the first couple of days after hatching. By the eleventh day, the end of the nestling period, they carry away all but 5 percent of the chicks' waste.

While this practice may seem bizarre and distasteful to humans, it is beneficial to birds because there is apparently enough undigested food in the sacs to provide nourishment to the parents, who during this period are often so busy feeding the brood that they neglect themselves.

In some species the method of waste removal changes as the chicks become older and stronger. Once young swallows grow strong enough, they back up to the edge of the nest to defecate. Pied Flycatcher nestlings hold the excrement in their bills and pass it through holes in the nest to their parents' waiting bills. Parent tits and starlings become less involved in removing waste once the nestlings have their feathers and are able to regulate their body temperature.

Some birds, especially those that eat lots of fruit, do not carry waste away from the nest. Generally these birds build nests in spots removed from predators. Many of these nestlings shoot the feces over the edge of the nest. Thus, while the nest remains relatively clean, the ground below is fouled.

Studies have shown to what length most birds will go to keep their nests clean. In many species nothing other than eggs or chicks is allowed in the nest. In one documented case, a nestling banded for observation had its leg broken by its parent, who was desperately trying to rid the chick of its band. In another incident, a parent American Robin fed its nestling a piece of meat too large for the chick to swallow immediately. The adult, in an attempt to "clean" the nest of this piece of food, threw it out, along with the nestling bird dangling at the other end.

While the majority of birds could be classified as "good housekeepers," some might, at best, be described as slobs. It isn't uncommon for the Great Horned Owl to accumulate so many dead animals in its nest that its young succumb to disease. The nest of a pigeon is a breeding ground for bacteria, mold, insects, and vermin. And after a visit to its dropping-encrusted nest, the parent kingfisher is so dirty that it has to take a bath.

Perhaps, though, the most blatant sign of neglect was documented in 1927 by E. A. Kitchin, who observed a large group of Pine Siskins, never a fastidious bird around the nest. A week later, upon his return to the woods, the birds were nowhere to be seen. What had become of the eggs, he wondered? He was appalled when he found out. All but a few of the nests in the forest were filled to the brim with excrement, under which lay the doomed eggs. (Obviously, this was a local phenomenon or the Pine Siskin would be extinct.)

LEAVING THE NEST

The fledging state – the time at which a bird leaves the nest – usually occurs as early as possible. As a general rule, the larger the species and the longer the incubation period, the longer the young stay in the nest.

Another factor that contributes to determining when a particular species leaves the nest is the risk of predators. Birds that nest in spots vulnerable to predation and other hazards are more likely to fledge early than those whose nests are less accessible. In some species, young birds are forced or encouraged to leave the nest because the parents are ready to resume egg laying.

The time at which a particular species fledges varies. In a study of eleven species of American open-nesting birds, for example, the average nestling period was eleven days, while the young of ten species of hole-nesters didn't leave the nest for nineteen days. Since most precocial birds leave the nest almost immediately after hatching, the term *fledgling* doesn't really apply but is used more properly to describe altricial young.

What compels a bird to leave the familiar confines of the nest and seek life beyond? It is believed that as birds mature, a certain instinct to leave takes hold. This may involve simply the sensing of an overfilled nest, if there are three or four large young within it. Often, however, the young require some coaxing from parents who are "eager" to get on with things. The nest, with its regular changeovers in incubation, feeding of the young, and consequent noise is apt to attract a predator sooner or later, so it makes sense to clear it as soon as the young are ready.

Usually this is done by reducing or even totally eliminating the amount of food brought to the nest in the final days of the nestling period. Timing, of course, is of the essence. If the food supply is cut off too early, the chicks may not be strong enough to survive; if they are fed too long, the adult birds may miss the chance to raise another brood.

Some species wean the young from the nest by dangling food just outside the chick's reach, requiring it to leave the nest if it is to eat. Bald Eagles literally starve their chicks out of the nest, and the Japanese Paradise Flycatcher has been observed rewarding two chicks who had left the nest with thirty feedings per hour, while totally ignoring the hungry pleas of the third chick, who was holding firm to the nest. Some birds of prey don't merely neglect their nest-clinging offspring, they harass them until – reluctantly – they budge.

Some species have been observed teaching the young while enticing them from the nest. A Belted Kingfisher, for example, was seen beating a fish until it was stunned, dropping it into the water, then watching while its young birds attempted to retrieve it. Similarly, Northern Harriers have been seen dropping mice in midair, with the young bird catching the prey with its talons after it had dropped only a few feet. One observer even noted a young Prairie Falcon practicing with a bit of cow manure, dropping it, swooping down, and picking it up, over and over again.

Flying exercises also are performed by some species prior to leaving the nest. Young eagles and hawks, for instance, stand on the edge of the nest and practice flapping their wings until one day they take flight. In some species this exercise appears to be essential, as is illustrated by Turkey Vultures confined in cages too small for wing movement. Even at three months of age, these birds were unable to fly.

The act of fledging does not necessarily end the parents' responsibility for the brood. In many species, the young birds would not survive if left to their own devices. Sometimes it is just a question of having adequate time for their basic instincts to emerge, along with some experience in the day-to-day world. Most species do, in fact, have some post-nestling care, even if it only involves parents leading the young to prime food sources.

Many young fledglings are not adept at flying, so they will perch somewhere near the nest, relying on the parents to bring their food until they are able to fly better. Yet some Arctic shorebirds leave their young before they can fly well, making it incumbent upon the young birds to feed themselves, as well as to find their wintering grounds on their own.

Some seabirds migrate with their young, feeding them for up to six months. Other nonmigratory species allow the young birds to stay until the next breeding season, when they are chased away and forced to go out on their own.

Natural selection – survival of the fittest – is never so apparent as it is with these young fledglings. Only those who have learned their lessons well will survive on their own, and fly off to fulfill the destiny of their kind.

The Bird's Digestive System

A bird's lack of teeth and strong jaw muscles requires that food be passed down the digestive tract before any chemical digestion begins.

Many bird species – particularly grain-eaters such as pigeons – are equipped with a crop, a storage compartment at the end of the esophagus where the food waits until the stomach can accommodate it. In some birds, the food is partially digested in the crop, where it may stay for as long as a day.

All birds have two stomachs, the glandular stomach and the gizzard. In the glandular stomach, the food is attacked by enzymes and acid, which break down the protein contents. Easily digested food may bypass the gizzard, an organ equipped for grinding food down. Lined with horny plates and filled with grit swallowed by the bird, the gizzard also serves as a trap for the bones and other indigestible parts that birds, particularly carnivorous ones, may consume. Bones, fur, feathers, and fruit stones may be regurgitated – coughed back up – sometimes in the form of pellets. In some species, the gizzard is so essential that if a bird is deprived of grit (grit in the gizzard is worn away by the grinding and must constantly be replaced), it will lose weight and die.

In the intestine nutrients are absorbed into the bloodstream. The length of a bird's intestine varies somewhat, to an extent determined by the diet. At the end of the intestine, many birds have a pair of dead-end tubes, the caeca. The job of the caeca appears to be the absorbtion of water and digested proteins and especially the bacterial decomposition of cellulose in some fibrous foods. This waste, dark and moist, differs from the white excrement most people associate with birds.

The intestine ends in the cloaca, the common end chamber of the digestive, urinary, and reproductive tracts (which in mammals are more or less separate). The cloaca is composed of three parts: one receives excrement from the intestine, another receives the discharges from the kidneys (urine) and from the vas deferens (sperm) or oviduct (egg), and the third chamber stores the excrement until it is ejected from the bird.

The digestive system of a bird is remarkably efficient. A shrike can digest a mouse in three hours, while seeds from berries eaten by a Blackcap have been found in the bird's excrement twelve minutes later.

3

How Birds Learn

Birds are not simply automatons whose actions are totally governed by genetics. Consider the behavior of an English titmouse and the Jackdaw.

A tiny Titmouse perches on the rim of a milk bottle the milkman has just deposited on a doorstep. With its beak, the bird methodically pecks away at the foil seal, tearing at it, layer by layer, until it finally gives way, exposing an inch-thick plug of heavy cream. The bird then dips its beak and drinks until the cream is gone.

The Jackdaw, too, has demonstrated an aptitude for acquiring learned skills. Its ability to count, for example, has been tested in experiments. In one, a Jackdaw was trained to lift the lids from a row of boxes until it had picked out five pieces of food. The bird was then trained to return to its cage. On one occasion the bird returned to the cage with only four bits of food, having taken one piece from the first box, two from the second, and one from the third. The experimenter was about to record the failure in the log when the bird returned to the boxes. It then bowed once in front of the first box, twice in front of the second, once at the third, ignored the fourth, and took a piece of food from the fifth box.

That birds are indeed capable of intelligent behavior is a relatively new concept. As ornithologists have had greater opportunities to observe birds in their natural habitats, they have come to develop a better understanding of the ways in which instinctive and learned behavior complement each other.

Instinctive behavior is adequate in most situations but it cannot cope with the unexpected. That is where learning enters the picture. Both abilities are important to survival.

In the absence of learned behavior, a bird can die, as will a pigeon born with a misshapen bill. It will try to eat as though its bill were normal but will find that picking up the food is impossible. Unless the bird is caught and the defect repaired, it will starve because in the pigeon pecking is an instinctive behavior with no room for change.

Instinctive or unmodifiable behavior is responsible for much of what a bird does. This is especially true in smaller species, which generally have shorter life spans and consequently less time to learn. Learning, on the other hand, is the process whereby behavior is changed. Behavior may be totally unmodifiable or partly modifiable – even learned behavior may have an "instinctive" basis.

A newly hatched duckling struggles to scale a barrier in order to follow a decoy. The more effort the duckling must exert to keep pace with this mother figure (as with its real mother under natural conditions), the more strongly it imprints upon the decoy.

Most often, learning in birds is the animal's attempt to adapt its behavior to a change in its environment. Birds that feed on diverse foods, in diverse ways, often learn how to deal with these foods. The tits that open milk bottles learn this habit from other tits. The habit may die out, and indeed occurs in some places and not in others, depending on the experiences of the tits in the different populations.

INSTINCTS

A Muscovy Duck, responding to the distress cries of a Mallard duckling, fights off the predator. But then it turns on the duckling and kills it in an instinctive response to the young duck's color.

A territorial, adult European Robin sees a bunch of red feathers and attacks it, ignoring a model robin that is perfect in all details save that it lacks a red breast. Again, instinct rules.

An adult Sooty Tern wildly attempts to drive a stray tern chick away from its nest. The chick struggles to avoid the jabs of the adult and, in the confusion, happens to touch the bird's chest. The action causes the adult bird suddenly to adopt the stray into its nest and care for it like one of its own.

Instincts – the automatic responses that characterize a species and that are shared by all members of that species – don't always make sense to human beings scrutinizing them under the microscope of reason. We don't perceive certain behaviors as log-ical: Why would a duck defend a helpless duckling only to kill it later? Why does a robin attack any red feather it encounters?

In studying the behavior of birds (or any other animals, for that matter), we humans must take care to avoid anthropomorphism, in which we invoke human motives to explain an animal's behavior. What we may fail to grasp is that a bird's senses give it a perception of the world entirely different from the one our senses give us. Birds react to things we are not conscious of; they may ignore what we assume to be obvious.

Thus, the robin that automatically attacks red is reacting to a stimulus, ingrained in its genetic makeup through generations

Examples of complex, learned behavior: First, the Cedar Waxwing closest to food (top left) will pass a piece to a less well-placed bird; and (bottom, left to right) a European Goldfinch uses its feet to lift food to its beak.

of natural selection. It may seem ridiculous to us, but the bird's automatic response to red is just one of the defenses that have enabled it to survive generation after generation.

A similar response is seen in young Kiskadees and motmots who avoid wooden rods painted with yellow and red rings. They recognize the pattern of colors as that of the coral snake, a feared predator. This is not a situation where the bird has the time or the inclination to discover through trial and error whether the colorful pattern it sees belongs to a snake or something benign. Thus, its instinct directs it; the room for error is too great, the stakes too high.

Many bird behaviors, including body care, feeding, fighting, aspects of nest-building, and of courtship, are thought to be largely unmodifiable or innate behaviors, although they may occur only at certain times, and in certain individuals, under the influence of hormones, only during good health.

Take bathing. Most bird species that bathe in water (others are dust bathers) immerse the head, suddenly raise it, and then beat the wings. This same set of motions has been observed in young birds who have no access to water. In the same way, species of hawks reared in captivity have been seen trying to take a bath on a piece of plastic food wrap. And month-old Northern Goshawks sitting on the bare ground have been known to go through bathing motions when they see another bird in the water.

The young altricial bird is able to do little more at birth than open its mouth: While newborn human babies instinctively know how to suck, newly hatched altricial young gape. Depending upon the species, a variety of stimuli will elicit the instinctive gaping response. Many young birds are equipped with swollen flanges at the angles of their jaws. These hotbeds of nerves cause the jaws to snap open at the least provocation, such as a slight shake of the nest or the parent's high-pitched call.

To ensure that the parents' desire to feed is as strong as the young's desire to eat, many species have mouth linings that are colored so as to arouse and direct the feeding instincts of the parents. Parrot-finches have five symmetrical spots on the roof of the mouth, a black bar crossing the tongue, and three bead-like opalescent emerald green and blue pearls at each angle of the jaw. Like a lighthouse beacon directing a ship on a moonless night, these pearls enable the parents to find the small mouths in the darkness of the covered nest, and to aim the food accurately.

The importance of such markers was documented in a study of Zebra Finches. A group of mutants born without any mouth markings was brooded in a nest containing chicks who had the markings. Adult bird mutants and normally marked parent birds took turns feeding. In all cases, the young birds who had mouth markings were fed first, received more food, and grew faster than the mutant group. Naturally, these birds had a higher survival rate.

Instinctive behavior in birds is often "particulate," in that the action is automatic and independent of anything else. The Sooty Tern who at first tries to drive the stray tern chick away from the nest, only to adopt it after it touches her on the breast, is a creature of the moment, reacting instinctively to each separate action. When she attempts to drive the chick away, she is responding to a strange creature who has invaded her nesting area. The accidental touch on the breast elicits another response, one that seems to be a direct contradiction of the original action. Behaviors are never wholly independent of anything else – they vary depending upon hormonal levels, arousal state, health, age, and so on.

To reach the food on top of the post, the Canary must perch on the toy truck, which is pulled by a thread. Even when the truck is hidden behind a screen, the bird has learned to pull the thread the number of tugs required to position the truck beside the post.

One widely accepted hypothesis that attempts to explain instinctive behavior posits the existence of fixed action patterns. These are rigid, stereotyped, predictable, and species-characteristic movements that are automatic. Independent of experience, fixed action patterns are complex actions that normally are precipitated by a simple stimulus. The first time an animal encounters the stimulus, it exhibits the fixed action pattern behavior. Yet both the context in which the stimulus appears and the motivational state of the bird are important.

The reaction of a female Herring Gull to a Herring Gull egg illustrates the fixed action pattern. The stimuli associated with an egg will incite feeding behavior if the female is not breeding and not incubating her own eggs, or if the egg is well outside her own nest. If the egg is *in* her nest along with at least one other egg, the female's incubation behavior will be released, provided she has not incubated the clutch for a while. That same egg, however, will elicit retrieval behavior if it is placed just *outside* her nest – and she will roll the egg back to within the nest's confines.

A bird's sensitivity to stimuli may change with time or as the bird grows older. When a gull finds a cracked egg in its nest early in the incubation process, it usually nibbles at the cracks, killing the embryonic chick within. But when the egg is discovered toward the end of the incubation period, a pattern of brooding and feeding is released.

In a famous study of Jackdaws, it was discovered that a group of tame birds would furiously attack anything or anyone dangling a black object. The experimenter first discovered this response when he took a pair of black swim trunks out of his pocket and was immediately swarmed upon by a black cloud of raging Jackdaws, attempting to peck his offending hand to pieces. The birds were normally on friendly terms with the experimenter, even allowing him to hold their unfeathered young in his hand without seeming concerned. However, once the young birds' quill feathers had opened, turning the birds black, he was immediately attacked by the parents whenever he held a baby bird.

Dangling the black bathing suit, it appeared, triggered the same sort of attack that the presence of a captured Jackdaw in a predator's mouth would incite. The birds were unable to differentiate between the two – and, of course, without a human presence, there would be no need to do so.

In some bird species certain behavior patterns have developed into exaggerated displays whose purpose is to transmit particular stimuli to other animals. Usually, the display acts as a releaser that precipitates a response, although sometimes it may serve to inhibit any action at all. Whether visual, auditory, chemical, or tactile, these releasers are particularly important in social communication between birds. Under breeding conditions, a female may crouch low and wiggle her wings to invite copulation by her mate. A similar behavior may also serve, in subordinate males as well as in females, to appease a dominant bird, preventing it from harming the displaying bird.

Not all behavior can be easily classified as either instinctive or learned. In many cases the lines between the two are blurred.

Most learned behavior does, in fact, have some unmodifiable elements, and some supposedly unmodifiable behavior can have some learning associated with it.

Take, for instance, a bird's ability to fly. Is flying a matter of instinct, something a bird's body is specifically designed to do? Or is it an acquired skill that a bird learns through experience and lots of practice?

If you've ever had the opportunity to watch a brood of young birds flapping their wings in the nest, as though practicing for their maiden flight, you will probably deduce that flying is something birds learn. How is it, then, that birds raised in captivity, in cages so small that opening their wings was impossible, can go on to soar across the skies, crossing thousands of miles of oceans along with other birds whose youth was spent entirely in the wild?

The Woodpecker Finch of the Galapagos Islands, unlike woodpeckers, does not have the appropriate bill and long tongue needed to dig into and probe for insects in tree crevices. So it has learned to find and use a cactus spine as a tool with which to dig out what would otherwise be an inaccessible food source. Young finches probably learn this habit by observing their parents, and thus by learning the technique it passes from generation to generation.

The answer is probably that flying is a combination of both unmodifiable and modifiable behavior. The bird is born with the ability to fly; but perhaps control, which enables a bird to land and maneuver so effortlessly, may be learned through practice. Physical and physiological development are also important – a nestling will not flap its wings in "practice" until the wing muscles are fully developed and its plumage is complete.

There are many examples in which a bird's instinctive pattern of behavior is modified to make it more efficient. The mature Loggerhead Shrike impales its food on thorns or wedges small pieces of food in the forks of twigs. A young shrike will hold the food in its bill and drag it randomly along the perch. If the food just so happens to catch on a thorn or wedge, the bird immediately becomes engrossed with repeating the action. Before long the bird knows how to direct the dragging action so that it achieves the desired result. Hence, the instinct to drag prey along the perch is refined by trial and error. Also, many predatory birds, as they become familiar with the habitat and prey will tend to concentrate on certain kinds of prey, perfecting their hunting techniques.

Chickens are born with the instinct to peck at small objects that contrast with the ground. As you can imagine, this indiscriminate pecking means that a chick may end up with a mouthful of grain, or it can just as easily come away with its mouth stuffed with pebbles or something equally inedible. Thus, it is incumbent upon the chick (usually under the tutelage of its mother) to learn to modify this behavior so that eventually it learns what to peck and what to avoid. Of course, chickens do eat small stones or grit, but even so, the young birds will learn to distinguish between appropriately and inappropriately sized and shaped stones. A combination of modifiable and unmodifiable behaviors helps many birds master the survival techniques they require.

IMPRINTING

A young duck hatches in a laboratory, sees the lab assistant, a bespectacled, bearded graduate student, and "adopts" him as its mother, following him everywhere. Obviously, it is vital for the duckling to identify its mother and, under natural conditions, the mother will always be the first large moving object seen by the just-hatched duckling.

So this seemingly strange response is quite normal. Filial imprinting, as it is called, is the tendency for young precocial birds to learn to recognize and then to follow their parents or, in some cases, substitutes. Usually, the substitute is a human who just happened to be the first moving thing the chick noticed on hatching but there are documented cases of birds imprinting on mechanical toys, shadows, moving boxes containing ticking clocks, and even a tractor.

Imprinting is critical to survival in precocial species because these birds typically leave the nest immediately after hatching, when they are led off by their mother to feed. A duck, goose, or chicken that cannot differentiate its mother from an inanimate or inappropriate object has little chance of survival.

How does a young chick or duckling distinguish its mother from other members of the species? The actual appearance of the parent is one distinguishing factor. Experiments have shown that chicks followed the hen that looked most like their mother, on whom they had imprinted.

In addition to visual recognition, the calls of the mother bird facilitate imprinting. While still in its shell, the Wood Duck embryo will respond to a Wood Duck's maternal call by increased bill-clapping, but it will decrease its clapping when a Mallard calls. The Wood Duck mother seems to sense that her calls are helpful in imprinting her young, so she begins calling every five seconds when the ducklings begin to pip at their shells on their road to freedom. By the time the ducks are almost hatched, she is calling out to them five times a second. The fine-tuning of the imprinting is absolutely crucial in some social birds because adults other than the parents may vigorously attack and could kill the young bird that approaches them.

Unlike most other kinds of learning, imprinting can occur only during a specific period. To find out more about that period, researchers exposed newly hatched Mallard ducklings to a mechanically operated decoy duck that had been painted to resemble a male Mallard. Each duckling was given a ten-minute run behind the decoy. The duckling was then taken away and later returned to test the strength of the imprinting. This was done by placing the duckling between two decoys – the original "male" and one painted to look like a female. If the duckling followed the male, imprinting was assumed to have been successful.

Researchers found that most ducklings imprinted between thirteen and sixteen hours after hatching. As a duckling aged, its ability to imprint markedly dropped. 80 percent of one-day-old ducklings not only failed to imprint but actually showed signs of fear, which made them avoid rather than follow the decoy. By thirty hours, none imprinted.

This short imprinting period is important because it keeps the young chick or duck from automatically following enemies or other ducks that might try to harm them, one of which the young birds are likely to encounter within the first few days of life.

While filial imprinting occurs only in precocial species, some species of both altricial and precocial birds sexually imprint – that is, they learn to identify characteristics of the parent or parent substitute that will later influence their mating preferences. This is important in keeping the species pure.

The degree to which a bird sexually imprints varies considerably. Canada Geese imprint easily, the Ring-necked Pheasant poorly. Some birds never sexually imprint. Imagine the problem a brood parasite such as the Cuckoo would have if it did. The young cuckoo whose mother laid her eggs in another bird's nest, only to abandon them, cannot afford to sexually imprint on the foster bird that broods it, or the cuckoo species would soon become extinct. On the other hand, there is evidence that a female cuckoo will tend to lay her eggs in the nests of the species that fostered her.

This period of sexual sensitivity occurs later than filial imprinting and is variable. The age at which the Mourning Dove sexually imprints is between the seventh and fifty-second days of life, while the Bullfinch is capable of sexually imprinting up to two years of age.

Ideally, American Robins recognize other American Robins, mate, and produce more robins. But what happens if a bird doesn't recognize its own kind? In captive situations in which one species of bird is reared by foster parents of another species, the young bird may sexually imprint on its foster parents. When it is mature, the bird tends to seek out mates that belong to its foster parents' species, ignoring potential mates of its own kind.

In experiments with sixty-eight male Zebra Finches raised by White-backed Finches (or Munias) and seventeen male White-backed Finches raised by Zebra Finch foster parents, researchers found that all the males directed their courtship toward females of the foster parent species. Females of their own kind were ignored and sometimes even attacked. Interestingly, however, when a Zebra Finch male was raised by mixed parents (one Zebra Finch and one White-backed Finch), he nearly always preferred to mate with a female of his own kind.

LEARNING

Alex is an African Gray Parrot with a vocabulary of some forty words. One school of thought holds that parrots and other talking birds merely mimic and do not relate words to objects. But many parrot owners disagree. When sprayed with water, Alex says "shower." Say the word "color," and he picks up a colored plaque rather than a plain one.

As one can imagine, studying learning in wild birds is not an easy task. Thus, most studies of this sort have been done with captive birds such as Alex. Often reports of bird intelligence are anecdotal, recited by proud owners whose objectivity is in question. And, fundamentally, "intelligence" is human-based, and use of the term with regard to other animals is difficult.

One accepted generalization is that the degree of intelligence in a bird's life relates to its needs, as well as to its size and the length of its life. Yes, the Jackdaw in experiments has been taught to perceive numbers and realize its mistake when it comes away from five boxes with only four food morsels. But is counting something the bird needs to survive in its world? Probably not. And can it take the information it has learned and apply it in a new context? Again, probably not. Many scientists, in fact, have wondered whether a bird counts its eggs or chicks and realizes if one is missing. But observations have been made of birds that will continue to sit on an empty nest, deserting it only after time, presumably when the lack of stimulus of the eggs against the brood patch causes the incubating urge to dissipate. In many birds, the clutch size is relatively fixed and the female ceases to lay when she feels the appropriate number of eggs under her; it is likely that learning is not a factor here.

Adaptive behavior in birds is a response to environmental factors that in some way threaten survival, either of the individual or of the species itself. Some birds have learned to recognize their eggs. These are birds whose nests are often used by parasitic species (who, like the cuckoo, lay their eggs in another bird's nest, and depend upon the substitute parents to brood and care for their young, often to the detriment of the bird's own brood). Many of these susceptible hosts have learned to recognize their own eggs as a means of survival. The African Village Weaver that sees a foreign egg in its nest will toss it over the side; some other birds will abandon the nest altogether.

The degree to which a bird can change its behavior is sometimes considered a measure of its "intelligence." These changes can only occur when information is stored in the nervous system and be called up when similar behavior is necessary to cope with a similar environmental hurdle. Species that feed in diverse ways or on diverse foods, or that can exist in different habitats, tend to have more open-ended learning potential, because their behavior requires flexibility.

Associative learning is one way in which birds adapt their behavior. This form of learning is accomplished by observing another bird, often a parent or older bird. Oystercatchers, for example, feed on mussels. Some individuals pry open each shell, while others hammer a hole through the shell. Each oystercatcher uses one method exclusively. In experiments in which eggs were moved from one nest to another, it was found that the chick's preferred method of feeding was learned from the foster parent.

The cream-stealing titmice at the start of this chapter used their beaks to poke through metal bottle caps in much the way that they would search for insects, seeds, and fruits in the wild. In fact, birds such as tits, Jackdaws, crows, and jays regularly seek food by probing, looking over and under branches, pecking as a test, and other exploratory ways. The milk-bottle context, however, is new, suggesting intelligent behavior. Not long after the

This drawing of a woodpecker's skull reveals the anatomy of its long tongue and the hyoid apparatus (u-shaped bone at the base of the tongue). Muscles control the protrusion and retraction of the tongue, which may wind over the skull to the nostrils.

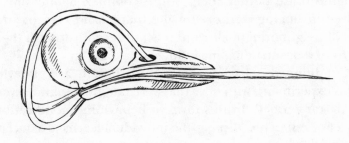

titmice began their cream sampling, other species of tits, as well as other bird species, learned this method, another example of learning by association. Furthermore, in some areas where glass bottles have been replaced by milk cartons, the birds have adapted their skill and become adept at opening the new containers, something with which some humans still have trouble. Throughout Great Britain, local titmouse populations have over time acquired and subsequently lost the habit of opening milk bottles, often later to regain the habit from a fresh initiator bird. This clearly shows the importance of learning to the titmice.

To test associative learning in Blue Jays, researchers placed the birds in separate cages, with one bird as student, the other as teacher. Some students then watched their teachers attack and eat butterflies of species A and others consume species B, which was of a different size and color.

The students were then offered a choice between a butterfly from species A or one from species B. Most of the students who had watched teachers devour A butterflies chose the same, while the ones whose teachers ate B butterflies also opted for their teacher's choice. One needs to bear in mind the rapid and very effective use of visual stimuli by birds – the wing pattern of a butterfly can be assimilated or acquired with relatively few observations.

There are different levels of learning; some simple, some more complex:

Habituation. In some instances, a bird must learn *not* to respond to a given stimulus. This is the simplest form of learning. Birds who live near freeways, for example, soon learn not to let the noise bother them. More complex habituation is involved in singing territorial males that learn to "recognize" songs of neighboring males and react differently toward them than to new or intruding male singers.

Conditioned Behavior. Most people have heard of Pavlov and his experiments with dogs that were taught to salivate every time he rang a bell. In this form of behavioral modification, known as *classical conditioning*, the individual learns to associate two stimuli.

In one behavior-modification experiment, caged Jackdaws were shown a ten-second burst of light, which continued along with four minutes of a recorded Jackdaw distress call. After each sequence, there were two minutes of stillness, with neither light nor sound. The Jackdaws soon learned to peck a key that turned off both light and sound.

Trial-and-Error Learning. Young chickens were offered colored water containing a substance that made them sick. One hour later they were offered the same tainted water. Not one chicken accepted a drink.

In *operant conditioning*, as trial-and-error learning is also termed, an animal learns to modify its response to a stimulus (or to make a new one). Through experimentation, the creature learns to repeat actions that elicit some type of reward, and reject those that are neutral or negative.

The ability of White Leghorn Chickens, Bobwhite Quail, Yellow-headed Amazon Parrots, and Red-billed Blue Magpies to learn through trial and error was tested in experiments to see whether the birds could learn right from left. Two containers, one empty, one with hidden food, were placed in front of the bird. Once the bird determined which one held the food, the food was shifted and the bird had to try again. After twenty-nine tries with successive right-left, left-right reversals, the results showed that magpies and parrots markedly outperformed chickens and quails.

Insight Learning. A Green-backed Heron spreads bits of bread in a pond and patiently waits, driving away other birds that attempt to make a meal of the crumbs. Moments later a school of fish, drawn by the bread, surfaces. The heron strikes, its patience rewarded by a good dinner. Herons and other fish-eating birds are aware of and watch other fish-eaters at a distance, and readily shift their foraging site when they observe a burst of successful fishing across a pond.

Insight learning, a higher form of behavior than those previously discussed, occurs when a creature, presented with a problem, suddenly spies a solution, and does whatever it takes to resolve the problem.

Springtime in Norway and Sweden finds some Hooded Crows pulling up lines that fisherman have poked through holes in the ice. A crow will grip the line in its beak, walk slowly backward as far as it can, and then walk toward the hole on top of the line. This prevents the line from slipping back into the water, and is thought to be an example of insight learning. Once again at the hole, the crow repeats the process until the end of the line emerges along with, in some cases, fish or bait. Unlike most birds, crows often spend time watching human activities, so it is not surprising that they are "insightful."

Some bird species have been shown to use abstract concepts in directing their behavior.

In a series of experiments, Canaries were taught to discriminate a unique object from other, identical objects. Nine objects were placed in depressions in front of the birds. Eight depressions contained aspirin tablets, one a wood screw. A piece of food was hidden under the screw, and the birds were trained to push aside the objects until they found the food. In the second trial, the birds were faced with eight depressions containing wood screws and one with an aspirin tablet, which this time hid the food. After 160 trials, the average Canary learned to choose the unique object fifteen out of twenty times. In later trials, the objects themselves were replaced by unrelated items. Performance improved with each trial.

Most documented examples of insight learning come from the laboratory rather than the bird's own environment. Studying most insight learning in a bird's natural habitat is difficult. One way, however, in which birds in the wild appear to use insight learning is through the use of tools.

While tool use among birds was once thought to be rare, at least thirty species of birds have been known to manipulate inanimate objects that enable them to perform a task more efficiently, whether that be building a nest or getting food.

The Woodpecker Finch of the Galapagos Islands is such a tool-user. This bird preys on insects that hide in tree crevices. Unlike the woodpecker, which has the perfect long tongue for probing the crevices, the finch is not so endowed, a potentially disastrous problem, considering these are the bird's major hunting grounds. To adapt for nature's shortcoming and avoid competing with a number of other species, the finch has learned to use a cactus spine to pry insects out of the tree wood. Upon finding an insect, the finch picks up a broken spine if one is nearby, otherwise it snaps a spine off the cactus. If the spine or twig has leaves or branches, the bird carefully trims them off to make the tool easier to manipulate. It then holds the spine in its beak and probes the crevice until the grub crawls out. The finch then drops the tool and gobbles up the insect.

To find food, some Brown-headed Nuthatches use scales of bark to pry other bits of bark loose. Occasionally, the Varied Sitella, an Australian bird, uses small twigs to probe for grubs in eucalyptus trees.

Rocks also are used as tools by some species. Black-breasted Buzzards in Australia have been observed smashing Emu eggs by raining stones on them. Similarly, the Egyptian Vulture will pick up a stone in its bill, which it throws at the egg of an Ostrich. Some scientists surmise that this action originally started with the vulture throwing small eggs to the ground to break them. Faced with the gigantic Ostrich egg, the vulture sought a different way to crack it. A stone was thrown, it just happened to hit its target, and the bird came to associate stone-throwing with food. Other Egyptian vultures observed the technique and, through associative learning, they, too, found a method to get at the delectable Ostrich egg.

If you have any doubt that birds are capable of problem solving, consider the curious crow family, whose members are considered among the most intelligent of birds.

A captive Blue Jay was faced with a dilemma. There was a pile of food pellets outside its cage, beyond the range of its bill. The bird studied the situation for a few moments, then proceeded to tear a piece of newspaper that lined the bottom of its cage. The bird poked the piece of paper through the wires of the cage and, using it as a scoop, lifted the pellets inside the cage.

As Blue Jays have never been known as tool-users, it was assumed that the bird's good luck was simply an accident. Even so, five nearby birds copied the technique.

Clearly, the ability of birds – Blue Jays and others – to adapt involves both instinctive and learned behaviors. As a bird matures, it employs both instinct and various modifiable behaviors in mastering feeding, fighting, courting, and nesting, just as the members of its species have for countless generations. All these are complex activities in which experience helps to mold the behavior. The act of eating an insect may be viewed as simple, but birds eat a wide range of insects that live very different lives, and no one method of catching the prey is adequate in all situations.

In the avian world there are other modes of learning, too, that allow birds to acquire skills that we are only beginning to understand and identify.

Anting

A crow spreads its wings, ruffles its plumage, and sits down on an anthill, allowing the angry ants to march through the forest of its feathers.

An oriole grabs one or more ants in its beak and jabs them into its plumage, concentrating on its critical wing and tail feathers. In the attempt to place the ants under its back feathers, the bird sometimes steps on its own tail and falls over backwards, a rather comical exhibition.

Both cases – the first passive, the second active – are examples of *anting*, a puzzling but widespread practice among birds. Between two hundred and two hundred fifty species of birds have been known to treat their feathers with ants. After the treatment, which may last up to forty-five minutes, the bird either eats the ants or throws them back on the ground.

Ornithologists are not certain why so many birds follow this practice, although they agree that it appears to be a largely unmodifiable behavior.

One explanation of anting is that the ants' formic acid may kill or dislodge lice from the feathers. In one experiment, a starling offered a boiled, acid-free ant, anted once but refused a second time. The bird did, however, ant several times when given dead ants whose acid had not been removed.

Another theory is that anting destroys feather mites. One researcher found that feather mites feed on lipids found in feathers and speculated that anting may reduce the amount of lipids, thereby starving the mites.

Other theories suggest that ants eat or repel a bird's external parasites, that anting relieves itching, or that it simply "feels good" to the bird. Moreover, some parts of a bird's body are not easily cleaned by the feet or bill; anting may help.

Whatever the reason, for many birds anting seems to be an addiction. Some birds, in the absence of the real thing, have been known to ant with beetles, wasps, or other bugs. Even more bizarre is to observe a bird "anting" with a mothball, a cigarette butt, an orange peel, a burning match, or beakfuls of smoke.

4

The Social Life of a Bird

The Jungle Babbler has been described as "awkward and gawky." While it may lack apparent grace from a physical standpoint, these small birds form a complex and stable society.

In the forests, woods, ravines, and even private backyards of India, Pakistan, and Nepal, the Jungle Babbler lives its life in small groups, usually of less than twenty. Unlike the young of most species, babbler offspring stay with the group for at least eighteen months. Those who do leave at this time are generally female, with male birds staying up to four years (less if a prime male dies in a nearby group) before venturing out on their own.

There is one breeding pair per community; with the help of other group members, they build an open nest, where the female lays three to five eggs. Like most birds, Jungle Babblers don't use the nest for sleeping. Rather, the members of the community line up, side by side, on the twigs or branches of a tree. At the center sit the breeding pair; two nonbreeding males flank the entire group, one at each end of the line, with the fledglings squeezed into the middle of the row, where the older birds' bodies can provide warmth. As the young birds grow, they gradually assume roosting positions toward the outer end of the line, but the fledglings are never allowed at the flanks, bastions strictly reserved for second-year or adult males. The world of the Jungle Babbler is an enviable one, in which order is maintained by strong leadership and the bonds between members of the society.

Most social bird species have a pecking order, a social stratification system in which the dominant members are the "kings" or "queens" and the others, to varying degrees, the subordinates. These systems are based on the health, age, size, and hormonal levels of the various individuals. Dominant birds usually are the first to feed, drink, bathe, and do everything else, while the subordinates are weaker, younger, or smaller birds who wait their turn. It seems to us, though, as if the Jungle Babblers' society is built around merit. The highest-ranking members in this society are the breeding pair, which appear to be the most competent, the hardest working, and are usually the oldest.

When clumped together on their perch, Jungle Babblers often preen or groom their companions. The two breeding adults of the community are the most active preeners and their efforts are most often directed toward the younger birds, most of which are their progeny.

A breeding adult – more often the male – is the leader on foraging trips, followed by other adults, and finally the younger birds. While the group forages, a sentinel perches several yards away, keeping watch for predators. After a while, the sentinel may be relieved by another bird who assumes the task. In five out of six cases, the sentinel is a member of the breeding pair; in other species, the sentinel is an experienced adult, but not one of the pair.

The main aggression in Jungle Babbler society is displayed by the younger members, presumably because they have to secure their place in the group. Young babblers have been observed "playing" by staging mock fights, displays in which a bird lies on the ground while others roll on top of it or gently peck at it. During their first year, babblers may be seen chasing one another, fighting over food, and generally disrupting the group.

Yet within a few months these birds have left their wild ways behind and become cooperative adults. Overt conflict ordinarily does not occur among adult members of a Jungle Babbler community. Whether accepting the succession of a new leader, should the breeding male die, or sharing food during times of scarcity, these strange birds have learned through centuries of evolution how to get along with members of their group.

❦

Chimney Swifts roosting in a dense cluster group during bad weather. This allows them to conserve heat and thus use less energy when food cannot be found.

Most birds do not belong to such a tight-knit community as that of the Jungle Babbler. Many of the birds we see in our backyards are rather solitary creatures who, except during the breeding season, spend most of their time alone.

There are, however, many birds who, at various times of the year, shun the solitary life and to some degree associate with other birds. Several species of birds join in roosting together.

A roost can be large or small. You may look out in your yard and see a single tree dotted with several dozen blackbirds settling down for the night. At the other extreme are roosts that have sheltered millions of birds. In the Mississippi Valley, winter roosts may contain as many as fifteen million blackbirds, while about five acres of groves near Lexington, Kentucky, were found to harbor several million roosting starlings, grackles, and cowbirds. If those numbers are not mind-boggling, just try to envision the winter of 1950–51 in Switzerland when an estimated 72 million Bramblings flocked nightly to the trees in two small pine forests.

These birds chose trees for their roosts, but that isn't always the case. During migration, Chimney Swifts have been known to roost in chimneys (they also nest in them), with several thousand clinging to the walls of one especially large chimney. Then there are the Common Goldeneyes who join with hundreds of birds to form avian "rafts," in which they then float along a lake or river. Swallows and weavers roost in reed beds, swirling down into them just before dark.

BIRDS OF A FEATHER . . .

Many of the birds who roost with others awaken with the dawn, fly off alone to hunt for food, and return to their communal roost only when the day is near its end. There are others, though, who stay with the flock during the day as well.

Some flocks are made up of one species; others are a melting pot of shapes, sizes, and colors. The common denominator is that the species in the flock usually can fly at about the same speed and feed in similar areas.

The majority of flocks are not permanent but come together at different times of the year. Some birds, an estimated one-tenth of the world's species, form colonies for the purpose of breeding. Other flocks may be temporarily drawn together after the discovery of a plentiful food source, only to disperse once that source has been exhausted. Others are seasonal, and still others are composed of migrants moving in a group. Herons feed both alone and, when the food supply is limited, in flocks. As the size of the flock increases, the number of fish caught per minute initially rises. After a while, however, the birds begin to get in each other's way. One observer noted that if a heron didn't catch anything after three and a half minutes, it gave up, and flew away. Thus, the size of the flock is regulated by the abundance of the food supply.

In a perfect world where food was always plentiful and enemies were few, it is doubtful that most birds (except the most sociable) would opt for life in a flock. On the surface it seems that the arrival of hundreds or thousands of birds at a feeding area would lessen each individual's chances of filling its stomach. Moreover, the pressure of living wing-to-wing can aggravate aggression, the result of which is more fighting.

Why do it, then? Because, once again, for every one of its disadvantages, flocking usually offers several advantages for many birds.

Take foraging for food. An Ostrich feeding alone has its head up on the lookout for enemies 35 percent of the time. When feeding with a companion, that rate drops to 21 percent, meaning the bird is able to spend 14 percent more time looking for food and eating. When three or four Ostriches feed together, each is able to spend 85 percent of the time concentrating on eating.

While it is true that some flocks require the presence of an abundant food source, birds that are part of a flock generally are more successful at foraging than their solitary counterparts. Studies have shown that chickadees flocking have much greater foraging success than a single chickadee, possibly because several pairs of eyes are better than one pair. Also, regular flock movements ensure that there is a time interval between feedings at any given site, so that no bird wastes its time feeding where others have recently fed.

Particularly in the northern winter, roosting en masse plays an important role in reducing heat loss. Two kinglets huddled together will reduce their heat loss by 25 percent. Add another kinglet to the equation and the loss of heat is reduced by a third. As noted earlier, fifty Long-tailed Tits were found curled into a feathery ball on a cold night. By huddling together, a large colony of Emperor Penguins, birds that inhabit the most frigid climate in the world, can raise the temperature at the center of their mass by as much as 20 degrees Fahrenheit.

Some groups can accomplish what the individual cannot. A flock of pelicans can encircle and trap schools of fish, something a single bird would be unable to do. The Cattle Egret uses a leapfrog approach in pursuing its insect food source. The flock rolls forward as the birds at the rear hop to the front. As

Masked Wood-swallows choose a secluded site and huddle close in the early morning in an attempt to reduce heat loss.

insects in front of the flock are chased, more are disturbed and moved, enabling the birds who bring up the rear to pick off their fair share.

Flocking birds also benefit from their neighbors' mistakes. Prey that is flushed and missed by one bird can be picked up by another. Ground hornbills, species found in Africa, walk in a line across fields, flushing and catching insects in a coordinated fashion.

Some flocks have highly organized feeding techniques. Blue-eyed Shags (or Cormorants) hunt in Antarctic waters in bird "rafts." The flock swims in close formation, with each bird periodically ducking its head underwater to look for fish. If one bird spots a fish, it dives under, followed by the rest of the raft. The advantage of this approach is that if one cormorant misses, another most likely will be successful. Such techniques require very favorable feeding sites.

Aside from foraging advantages, life in a flock is generally safer than living alone. Wintering shorebirds are preyed upon by the small hawks known as Merlins. A lone shorebird is three times more likely to be killed by this predator than is one belonging to a flock. Experiments have shown that the collective response to a predator is superior to that of an individual bird. A group of ten captive starlings were on the average one second quicker to react to a model hawk than was a single bird. One second may seem insignificant to us but in a bird's world it can mean the difference between living and dying.

Because of its communal eyes and ears, the flock is less vulnerable to surprise. Flock members warn each other of danger. Ducks flush together at the sound of a predator, with each member of the flock synchronizing its takeoff. The result is a collective movement, presenting a much more difficult and dangerous target than ducks flying at random. Shorebirds and starlings fly in dense flocks that are very maneuverable, thus remaining elusive and dangerous for predatory birds, which could be injured flying into a flock. A falcon traveling at 140 miles per hour can never "afford" to even touch a moving bird with its wing; the wing would break. Only at the right moment and only with its talons can the falcon strike at that speed.

The presence of a flock also enables birds to take a more active stand against a predator. A flock of birds sometimes will attack or mob a predator in an attempt to drive it away. A group of crows will mob a Great Horned Owl. A single crow wouldn't attempt this approach but the risk is markedly less in a group.

Some birds form tighter formations when a predator is nearby. Instead of trying to fly away, many flocks close up ranks and circle, or fly erratically in an attempt to confuse the attacker. When faced with a swirling mass of birds, some birds of prey back off, concentrating their efforts upon stragglers. The risk of injury from an imprecise high-speed strike is too great for these predators, which must attack with precision and maintain their bodies in good condition if they are to be successful.

SOCIAL FACILITATION

Watch a hen feeding at the farmyard trough. After a while, she becomes full and leaves the piled-up food, not to return again until she is hungry. But add another bird to the trough, and the first hen, the same one that only moments before couldn't eat another bite, returns to the food and proceeds to stuff herself, consuming on average 34 percent more grain. Add one more bird to the group and that first hen will eat 53 percent more than when she was eating by herself.

The way in which the behavior of one animal is enhanced by another is called *social facilitation*. Briefly, social facilitation is a contagious form of behavior in which one animal performing what is more or less instinctive behavior acts as a releaser for the same behavior in another individual.

In experiments to determine the effects young chicks have on one another, the chicks were deprived of food. After six hours without food, an individual chick made about forty food pecks during a five-minute interval. However, when the same chick was eating next to a nondeprived chick, the rate of pecks increased to 115. When paired with a chick that had also been deprived of food for six hours, each chick pecked 150 times per five minutes, the rate increasing to 260 pecks when the pairing was with a chick that had not eaten in 24 hours.

Like many behavior patterns, social facilitation doesn't always make sense nor is it always in the bird's best interests. A minor disturbance can cause a flock of Roseate Spoonbills to desert their nests en masse. In the Rook community, a little pilfering of sticks from each other's nests is common and tolerated. As long as the thefts are minor and contained, life goes on normally. Occasionally, however, the urge to steal reaches epidemic proportions, a problem that can occur when a rookery has only a few trees and sticks available. Then unabashed thievery may rapidly spread through the rookery, ultimately causing the wholesale destruction of nests.

Then there is the case of the hungry chick placed in a pen with a group of sated birds. However hungry it is, and however close to food, the bird will not eat because the others are not. The food is there, the bird is hungry, but because the others are not eating as they normally would, the bird senses danger and follows the action of the group.

A BIRD'S PERSONAL SPACE

Atop a roof, nine Herring Gulls perch, each so precisely spaced from the next that it is as though someone with a yardstick had marked off designated positions for each bird.

Whether hunting for food, flying, or simply perched on a telephone wire, most birds keep a set distance from one another. This individual distance varies from species to species. A Common Black-headed Gull will tolerate another bird standing more than a foot away, but should that distance be reduced to less than a foot, the offender is likely to receive a sharp jab or an equally hostile expression of displeasure. A heron needs even more space – several yards to call its own.

In one experiment, hoppers full of food were moved progressively closer to find the distance at which a male Chaffinch would tolerate another male. The Chaffinch allowed the male bird to approach until it was within eight inches of the hopper, at which time the bird behaved aggressively. Female birds were allowed to come within half that distance before the bird became aggressive.

While many birds keep this strict distance, there are those at the other end of the spectrum who prefer having another bird nearby, usually a mate or offspring and the closer the better. The small parrots known as lovebirds probably got their name because they are such sociable creatures, huddled shoulder to shoulder, packed into sardine-tight clusters, bathing together, grooming each other, even, occasionally, feeding each other's young.

It is not difficult to see the advantages of maintaining a certain distance from one's neighbors. Anyone who has maneuvered his or her way through a lunchtime crowd on a pedestrian-jammed city street has probably at one time or another been jostled or found himself on the receiving end of a sharp elbow. Close proximity can breed aggression in both the human and bird worlds, but, unlike us, birds usually manage to circumvent the hostility.

Keeping a safe distance from one another also has its practical advantages. We look to the sky and see a flock of birds flying as though it were one perfectly synchronized machine and not dozens of individuals. But however close they appear, each bird is flying at a certain distance, appropriate for that species, from the next one – a necessary practice if collisions are to be avoided.

Then there is the matter of feeding. At times Rooks feeding together must maintain a certain distance from each other so as not to give the earthworms on which they feed the advance warning that would enable them to escape down their burrows. This usually occurs early in the morning or after a rain, when the worms are nearest the surface and more apt to sense any movement. Sometimes the discovery of a particularly abundant food source is enough to make birds temporarily suspend their need for a set space, but it usually isn't long before tempers rise, squabbling occurs, and a more distant feeding pattern is resumed.

PECKING ORDER

A brood of newborn barnyard chicks is a fairly democratic group. For the first few weeks of life, the young chicks live in their pen quite peacefully. Gradually, though, they begin to peck and jump at each other, until, by the seventh week of life, a noticeable change occurs.

Suddenly it becomes clear that all chicks are not created equal. A social hierarchy has evolved, in which some chicks have already assumed higher positions on the ladder of success while others grovel, unable even to climb the first rung.

In the world of these chickens, there is room for only one top bird, the Alpha. This bird has earned the right to move and feed at will in the flock; it has achieved dominance. The Alpha is the bird most adept at getting what it wants, and is often the biggest and healthiest. If it wants to eat and another bird is in its way at the trough, it will peck the offender until the spot is vacated. If this dominant chick sees the warmest roost site occupied, it doesn't hesitate to oust the other bird, although the subordinate bird usually gets out of its way before it is pecked. In its first few weeks of life, the Alpha bird clearly demonstrates its dominance over the others; the chick emerges as the superior one in the flock, able to peck any other chick if necessary, but not having to do so very often.

The Alpha chicken, as you can imagine, leads a very good life, compared to the other birds. It gets the best and the most food, the most desirable roosting spot, and is usually the most successful breeder. The second-ranked bird, the Beta chick, doesn't have it too bad, either, because Beta has achieved dominance over every bird except Alpha. The next down, Gamma, can peck every bird except its betters, Alpha and Beta, and the hierarchy continues all the way down to poor Omega, the bird who can peck no one but who is pecked by everyone.

❦

A pecking order occurs only in permanent or semiperma-
nent flocks or social groups of birds. The birds learn to recognize each other and the order of precedence is settled early on,
although pecking order does change in wild birds as a result of
unstable health and hormonal levels in individuals. The advantage of a social organization in the flock appears to be a marked
reduction in conflict, which allows the individuals to spend time
in more constructive pursuits such as feeding.

In one experiment, two flocks of hens were compared to
see the effect of the absence of pecking order. One flock maintained a rigid order, while experimenters frequently moved the
other group's members, disrupting the order. The birds in the
disorganized group fought more, suffered more injuries, ate
less, and gained less weight than those in the organized flock.

Pecking orders vary, depending upon the species. Species
such as titmice, doves, and Canaries have a loosely organized
social hierarchy, while in the chicken and Jackdaw communities
the pecking order is rigid.

Some flocks of poultry are not organized in the linear manner described above but in triangles. Bird A pecks Bird B, which
pecks Bird C; but Bird C then pecks Bird A. In some flocks there
are one or two birds that dominate all the subjects equally.
Among Canada Geese, pairs with young are dominant over
those without. In this society, barren pairs are dominant over
their superiors' young, but single adult geese are subordinate to
the young as long as they still live with their parents.

Once a chicken has established its superiority, its authority
rarely is challenged. In this rigid system of pecking order, the
dominant bird is said to have pecking rights over its subordinates. All the dominant bird need do is make the slightest
threatening gesture and the subordinates instantly scatter. A
low-ranking Jackdaw in the presence of the top-ranking bird
bends its head low and turns the nape of its neck toward the
dominant bird, a gesture reminiscent of a subject bowing before
the king – but let the king dare to become ill and the bird will
be challenged repeatedly.

In some more loosely organized hierarchies, birds continue to jockey for dominance and the bird that wins the majority of these contests is said to have achieved dominance.

How, then, does a bird achieve this dominance? As a general rule, large birds dominate smaller ones and older birds dominate younger ones. The most belligerent birds are more likely to come out on top than their meeker counterparts. Sex is also a factor: Roosters usually dominate hens. But males and females in many species take turns being the dominant bird. In many finch species the males are dominant in the winter; but come spring, the females "rule the roost."

Hormones have been manipulated in some bird species in order to study the effects on social status. Results have varied but it appears that in many cases there is a correlation between social status and level of male hormones.

In one study, testosterone, a male hormone, was injected into low-status hens. Following the injections, the hens laid fewer eggs, but their combs grew and they began to crow. These hens then revolted against their superiors, and triumphed. Even after the treatment was discontinued, the hens remained dominant. Similarly, the injection of testosterone into male Chaffinches boosted their social status, a result duplicated in trials using female Canaries and doves of both sexes. Conversely, neither massive doses of male hormone nor even castration changed the social ranking of European Starlings.

In some species a bird achieves dominance without any signs of physical violence. Territory can be important in the social hierarchy, and any newcomer is subordinate, at least initially. In one experiment, Canaries were put into cages already occupied by other birds. The newcomers immediately assumed subordinate positions within the established community. In the pigeon world, even the most lowly will dominate a higher ranking bird that is new to an area.

Coloration in some species is related directly to ranking. Harris' Sparrows vary widely in the amount of black they have on their heads. Older males have a lot of black and are dominant over the young birds who have almost no black at all. In one experiment, the birds' appearances were altered to determine whether color would make a difference in status.

The heads of young birds were blackened with shoe polish and the older birds' heads were bleached. The result was a complete role reversal. The newly dominant young birds began to behave in a dominant manner. The older birds who suddenly had white heads found themselves forced into more combative situations than before, when they were the undisputed leaders of the flock. It wasn't long before they began to behave like subordinates.

Like humans, some birds can improve their lot by mating with a bird of higher status. One observer noted that, in a semi-wild flock of Jackdaws, life radically changed when a low-ranking female was chosen as a mate by a high-ranking male. The female, abused at one time by most of the flock, suddenly acquired "first lady" status and never again suffered physical indignities from any member of the group.

Some males have enjoyed similar upward mobility. In one instance, a Steller Jay that ranked in the bottom third of the flock's social hierarchy, eventually rose to become the leader after mating with the group's most dominant female.

The advantages of life at the top of the avian social ladder are many. Higher-ranking birds eat better than their low-ranking neighbors. The dominant hen gets priority at the feeding trough, with the lower-ranking members of the flock allowed only the leftovers. It isn't hard to understand why higher-ranking birds generally live longer than those at the bottom of the avian social ladder.

Another major benefit of rank is its relationship to breeding. Generally, higher social rank equates with greater breeding success. The highest-ranked rooster usually is in great demand as a breeder, while the low rooster on the proverbial totem pole is hard-pressed to find a willing mate. In many cases, the lowest-ranked male never breeds at all. Interestingly, even in experiments when every other rooster was removed from the flock, the low rooster still did not mate, an apparent victim of the role his society had thrust upon him.

It is worthy of note that, if food is readily available and sufficiently dispersed, all flock members benefit from not having to undergo combat. So a dominance hierarchy is not detrimental in itself. Given that feeding flocks of wild birds frequently are disturbed by predators, such a hierarchy allows a flock to move into a feeding area rapidly and food to be gathered by all without any fuss.

The explanations for the varied social arrangements in the bird world are numerous and more are constantly being discovered. There are pecking order and the instinct to flock, social facilitation, hormones, and the inbred knack of maintaining natural spacing. But however intriguing the explanations, the forces that guide the social lives of birds reveal much about each individual species and its interaction with the world around them.

The face of a Common Barn Owl is not perfectly symmetrical. The left ear is higher than
the right and is most sensitive to sounds from
below the horizontal. Conversely, the lower
right ear is most sensitive to sounds from
above the horizontal. Because of this asymme-
try, a sound will arrive at each ear at slightly
different times. This enables the owl to pin-
point the source of the sound to hunt successfully in total darkness. The ear opening is
covered by feathers on the sides of the head
behind the eyes.

5

Bird Communication

If the birds of the world were to have a vocal talent contest, one prizewinner might be the male Northern Mockingbird.

To look at him, you might not expect much. He lacks the brilliant plumage of the Sunbird, the regal demeanor of the swan, the flitting grace of the hummingbird, or the awe-inspiring power of the eagle. On the surface, he is quite ordinary, with his brownish-gray feathers, and small, plump-breasted body. But any impression of ordinariness fades the moment he opens his mouth.

It is then that the Northern Mockingbird shines. The bird not only has a beautiful voice that it generously shares with an appreciative world, but it is also an expert mimic.

Mimus polyglottos is the mockingbird's scientific name, which means "many-tongued mimic." Unlike parrots and mynahs, more renowned mimics who have only been found to copy speech patterns when they are in captivity, the mockingbird performs its feats of mimicry in the wild.

In between its own bursts of song, the mockingbird on any given day may imitate the songs of dozens of other bird species. One mockingbird was once observed imitating a species that lived hundreds of miles away. How it had learned the other's song is still a mystery. The tinkling of keys being played on a piano, the barking of a dog, the sound of an axe connecting with wood, a wolf whistle, and the honk of a car horn are some of the other sounds mockingbirds have been known to make. This small bird has a large song repertoire – some mockingbirds have been known to sing two hundred songs.

Why does the male mockingbird sing? Certainly to keep other males from his territory, which he vigorously defends. But it also appears that in this creature's world, the complexity of a male's song may be the yardstick by which a female measures a potential mate's worth. Thus, intricate song equates with a prime territory, capable of adequately supporting a brood of chicks.

Anyone who has taken a stroll through a forest in spring and listened to the symphony of birdsong might suppose that most birds are capable of uttering such sweet sounds. In fact, only a small proportion of the world's bird species are song virtuosos, at least as far as we humans are concerned.

If you were to face a singing bird, you would see a creature with its beak wide open, which is the normal method of singing for the majority of birds. Unlike humans who have a larynx and vocal cords, the bird has an organ called a *syrinx,* a sound-generating structure that is unique to birds, typically located at the lower end of the windpipe. One distinctive feature of a bird's voice is that it may produce as many as three or four complex sounds simultaneously. At the same time, the speed at which a message can be transmitted and received is perhaps ten times as fast as our own transmission system.

When it comes to what comes out of their mouths, birds do run the gamut from the nearly mute stork (which, however, can clap its bill quite loudly) to captive parrots that have been taught to recite their ABCs. Species of the order Passeriformes are those generally called *songbirds,* and to one degree or another are adept at singing. Thus, the Song Sparrow has nearly nine hundred variations of its song, while the Chipping Sparrow just repeats one song over and over. There are also many birds that can't carry a tune but utter calls as a form of communication.

BIRD CALLS

In the heart of Africa, tribesmen listen carefully for the calls of an African Greater Honeyguide. The bird, usually alone, seeks out humans, then gives a wavering, chattering call. If a human responds, the bird flies from one perch to another, calling as it goes, leading the person eventually to a nest of wild honeybees. Breaking open the nest, an impossible task for the small bird, is easily accomplished by a man or perhaps a large animal, possibly a honey badger or even a baboon. The honeyguide sits quietly, waiting for the sweet honey to be taken out, and then flies to the hive to partake of the bits of honeycomb left by the invader. Indeed, it is common practice for the human honey-hunter to leave a large piece of wax for the bird.

The honeyguide's chatter is but one example – albeit an unusual one – of a bird's use of verbal communication to help meet its needs; it may call for as long as half an hour, if it is being closely followed, before reaching the hive.

Verbal or acoustic communication among birds usually involves calls. By definition, calls are briefer than are songs. Birds have a great diversity of calls. The Australian birds known as miners have as many as ten alarm calls, each for a different type of predator.

Bird calls take many forms and serve many functions. Generally, calls are associated with finding food, avoiding enemies, parent-young relations, social contacts, and the movement of the group. Basically, a call transmits information to another individual, who then can behave accordingly. Because a calling bird attracts attention to itself, we can assume that the calls we hear have evolved to serve particular functions.

Both calls and songs appear to have instinctive bases. Critical calls that are immediately important to young birds usually require no learning; the young beg instinctively and react instinctively to adults' alarm calls.

Consider the just-hatched Greater Prairie Chicken. If the mother gives her "brirrb brirrb" call, the chick will come running to her. But if she gives another call, a shrill warning, that same chick will automatically freeze.

A mother hen emits a drawn-out squawk that warns her chicks of danger overhead from a bird of prey. The chicks, in turn, scurry for cover. If the mother were to make a series of rapid clucks instead of squawking, the warning would be of the approach of a ground predator such as a weasel or fox.

A bird who spots danger often will give a warning call to alert other birds and to send a message to the predator that its little surprise has failed. The alarm call not only communicates the bird's state of alertness to the predator, but causes movement and other calls that distract the enemy and may cause it to seek its meal elsewhere, especially since other species of birds in the area often add their alarm calls as well.

Among small birds there are two general categories of alarm or warning calls. The first, a "chat" call, is an abrupt, harsh warning. The caller's location is easy to find, making it prudent for a bird using this call to be sure it is a safe distance away from the danger. In contrast, "seet" calls are thin and high-pitched, with a narrow range of frequency that rises and falls, making it difficult for a predator to trace accurately the source of the call. A bird who spots a bird of prey flying overhead might take cover in the foliage and warn the other birds with a seet call. Of course, even a bird who utters a seet call is not immune to danger. Experiments have shown that in certain situations some birds of prey can locate the source of these calls.

The European Chaffinch will give a short, low, "chink-chink-chink" call when it sees a hawk perched in a tree. The call may attract other "chinking" small birds, who will then mob the hawk. If, however, the hawk is in flight, the Chaffinch sounds a seet call, flies into any nearby foliage, and repeats the warning.

In some species, a rallying call will bring birds of a kind together for a joint defense. When a Common Crow comes upon an owl, it cries out with many repeated "caw" calls, which will bring any crow within earshot flying to the spot to join it in mobbing the owl.

Experimenters in a forest saw not a trace of crows. However, when they broadcast an amplified version of the crow assembly call, the site was soon swarming with the birds. Moments later a recorded crow alarm call was broadcast, and the crows disappeared as quickly as they had materialized.

Another type of bird call is the contact call, a form of communication some birds use to keep others aware of their position. It is particularly common in social birds that live in a group, and in paired birds that feed somewhat apart on the territory.

A flock of finches will feed in the forest. After a while, one bird decides it is time to move to another spot, so it begins calling to the others, which usually stop feeding and take up the call, until all members of the flock have left. If one member is mistakenly left behind and becomes lost, it flies in search of the others, calling loudly and continuously until it meets up with them.

Sonograms of contact calls have shown them to be generally similar among some species. The calls tend to be short bursts, which span a wide range of frequencies. In essence, this similarity means that certain species are able to some degree to communicate with others, an advantage particularly in mixed-species flocks of birds. Unlike contact calls, which are used generally among species members, there are also soft "intimate" notes and courtship calls that are used only by adult males and females during and after pair formation.

Male Manx Shearwaters return to the colony at the start of the breeding season and move into nesting burrows, awaiting the return of their females. The arriving females call from the air, often in the dark, and the males answer from the deep underground burrows. Each male's individual call is like a beacon, lighting the way for the female, guiding her into the right burrow. Through learning (habituation) during the original forming of the pair, the birds know each other's individual call, even against a background of very similar (or to our ears identical) calls.

Similarly, birds who live packed tightly together in crowded colonies emit calls to their mates prior to landing in the nest, so they will not be mistaken for an intruding bird and attacked by the protective mate. Gannets' nests are spaced no farther apart than the range of the bird's long beak, so nest occupants are always on the lookout for intruders. This makes it essential for the returning bird to alert its mate to its impending descent. The tight spacing, of course, permits the maximum number of nests on the protective cliff face where Gannets typically build them.

Eighteen Common Ravens were raised in aviaries and allowed to fly anywhere within their confines. Members of a mated pair recognized each other and communicated information only to their partner, even when the mate was not nearby. When one raven was off flying, the mate frequently uttered sounds that normally were made exclusively by the absent bird, which, upon hearing the sound, would return immediately to its mate.

Many birds can recognize the songs of their neighbors; except for alarm calls, songs are usually the only loud calls that can be heard from all directions by a male on its territory. This is especially useful in setting defense priorities. A territorial male is less apt to show aggression against a neighbor that doesn't cross its property line than it is against a stranger.

Nowhere is instant recognition more important than in the relationship between parents and offspring. In some cases, recognition begins even before the chick is hatched. Common Murres build no nests but pack together on broken areas along cliff faces, potentially perilous spots in which to raise young birds. Thus, it is vital that families recognize each other from the onset. This is accomplished by the chicks calling to the parents from within the egg days before they are hatched, and the parents returning the calls. By the time a chick is hatched, it will respond only to its parents' calls, should it become separated from them on the ledge.

In the world of the Adélie Penguin, adults returning from the sea with food call to their young, who come running, calling back to the parents. Before any feeding takes place, the parents and chicks engage in a mutual calling display, seemingly to identify each other. Chicks who approach the wrong adults are driven away. Although it is possible that the parents also can identify their chicks by sight, chicks do not respond to silent adults.

In experiments where the young were separated from the parents, the adults were easily able to identify their own calling chicks. However, when the chicks' bills were taped shut, the parents were successful only five out of ten times. When an adult penguin adopted a mute chick, it cared for it only until it heard the call of its own chick. Likewise, in an experiment with chickens, an alarm-calling, wing-beating chick when covered with a thick glass bell jar was totally ignored by the mother, which responded to it instantly when the jar was lifted and she could hear, rather than just see, the chick.

The precise "vocabulary" of any bird varies. Ornithologists have distinguished ten different vocalizations, each with its own meaning, in the domestic chicken. Other birds tested include the Red Grouse, in which fifteen calls were identified; the Ring-necked Pheasant, sixteen; the House Sparrow, with eleven separate calls; and the Village Weaverbird, with fifteen. These calls include songs, which in some cases are variable, with several or more types, and occasionally hundreds of variants from a type.

NONVOCAL SOUNDS

Among some bird species the sounds of feathers drumming, wings flapping, or beaks hammering are significant communication signals.

The male Broad-tailed Hummingbird is one such bird. The outer feathers of his tail are modified so that they emit shrill whistling sounds during territorial display flights. How important is this? In experiments, males were altered so that their feathers caused no sound. The result was that the silenced males became less aggressive in defending their territories and ultimately lost them to males whose feathers had not been altered.

The White-bearded Manakin uses its wing feathers not in territorial displays but in courtship. The manakin's primary wing feathers are markedly narrowed and stiffened, allowing it to make snapping or growling sounds when the perched bird beats its wings. The sound – two clicks followed by what sounds like the buzz of an insect – appears to be part of courtship.

The European Common Snipe, on the other hand, uses some modified outer feathers on its tail, not on the wing, when it goes courting. As the bird takes to the sky in courtship flight, its unusually stiff and narrow tail feathers spread outward and vibrate, generating a sound that has been likened to the bleating of a goat.

In central Africa, the unusual sound made by the tail feathers of the Lyre-tailed Honeyguide is more easily identifiable to local residents than the bird itself. This bird, which has not one vocal song (but several calls that are used in other ways), nevertheless makes interesting music as it drops in spirals from hundreds of feet up in the air, the rush of the wind through its spiky feathers creating a sound that can be heard throughout the forest.

Some birds use their beaks to communicate. Storks, owls, shrikes, and many other birds clap their mandibles together, while woodpeckers use their special chisel-tipped beaks to hammer on dead trees or even on tin roofs.

Such hammerings, called drumming, have been shown to be important to the Great Spotted Woodpecker when courting. Normally, the male bird gives drumming bursts one hundred to two hundred times a day, but during courtship the number increases to between five and six hundred drummings; effectively, drumming is its courting song. Trees with their varied resources provide sound boxes for woodpeckers. By selecting certain hollow trees for drumming, the male woodpecker can communicate to prospective mates something about his environment, i.e., that he has a tree suitable for excavating a nest. Once the bird mates and begins to excavate a nest, he may go for several days without drumming.

WHEN A BIRD SINGS

The question is often asked, Why do birds sing? To a human enjoying a day in the spring woods, listening as the quiet is transformed by a veritable avian orchestra, it may seem that the sole purpose of a bird's song is to entertain us, the appreciative listeners.

Or, it has been written – in more places than it is possible to count – that birds sing because they're happy.

However reassuring it may be to think of birds as greeting card caricatures – plump, rosy, and smiling – it is doubtful that entertaining humans or expressing its feelings of happiness is the motivation behind a bird's song. In a sense, a singing bird is communicating that it is healthy, strong, and in possession of a territory.

The simple fact that songbirds spend so much time in song attests to its importance. In one observation, captive male Song Sparrows on a given day spent nine hours singing, nine hours sleeping, and six hours eating and pursuing other activities. A Red-eyed Vireo sang 22,197 songs in one 24-hour period, while a tropical manakin spent 86 percent of its waking hours in song.

Why then does a bird sing? It appears that the song of a bird has not one but many functions. Among them are the following:

Species Recognition. Every singing species has a specific song or songs or even parts of its song recognizable to other birds. A European Robin's song is complex and warbling, with each song burst made up of four different phrases. One robin alone has a repertoire of several hundred phrases, which it uses in various combinations. Although the songs vary individually, other robins and many humans, too, can identify a robin song, chiefly because the quality of the notes is similar, the songs are of a similar duration, and certain phrases occur regularly in all robins' songs.

In an experiment to see whether robins would recognize altered robin songs, tapes of artificial and natural songs were played to territorial male robins. The birds reacted toward the normal control songs by approaching the speaker aggressively. When the artificial songs were played, however, their reaction was milder.

Just as every species has its distinguishing songs, in many cases individuals have vocal variables that make them easily identifiable. A male bird on a territory, for example, quickly learns the identity of his neighbors by listening to their songs. This auditory recognition benefits the bird by enabling him to differentiate between a neighbor flying near the border of the adjacent territory and an intruding male that is considering an outright invasion. Playback experiments have shown that a territorial male responds with great agitation to songs of neighboring males played in its territory, but the greatest response and agitation comes when one plays back the male's *own* song!

Territorial Proclamation. Song is important both in establishment of a territory and in its defense. What to us sounds so sweet may, in fact, be a warning to other males: Stay away or else! Of course, a male without a territory sings less often, moves constantly, and is unlikely to breed.

To determine the importance of song in territorial defense, many experiments have involved removing males from their territories and replacing them with a sound system of broadcast bird song. When Great Tits were tested in such a manner, males searching for territory settled in the silent control areas hours before they invaded those with broadcasted songs.

Great Tits, like many birds, sing several songs, the average being three or four. Each song is sung several times before the bird goes on to the next one. It has been suggested that this may be a way of trying to deceive a potential invader into thinking there are more males present than is actually the case. Studies have shown that Great Tits with the largest song repertoires are more likely to survive for more than one breeding season and produce more young than those with smaller repertoires. In experiments where one area was broadcasted with one song repeatedly played and another with several songs, researchers found that invaders were quicker to occupy the territory from which only one song was being broadcast.

Some species like many white-eyes and tyrant flycatchers

have a very different, loud, and often repeated song they sing only at dawn, or at dawn and dusk. Quite often, another distinct song is sung during the rest of the day.

Oftentimes large territories correlate with large song repertoires. Male Red-winged Blackbirds that have the most impressive song repertoires also have the best territories, two factors that combine to give the birds the best breeding success. "Success" includes the male attracting not one but several females, which then nest in its territory. From an evolutionary standpoint, these successes ensure the further development of even more elaborate songs as males continue to compete for females.

Reproductive Functions. In most singing species, it is the male bird who does the singing, although in some species such as the Red-winged Blackbird, the Northern Cardinal, the Black-headed Grosbeak, and many tropical birds, the females also sing. Some female barbets and tropical bush-shrikes sing regularly with the males in very synchronized duets.

There is no doubt that male hormones stimulate singing in a variety of birds. If a male Chaffinch is castrated, it stops singing. But administer the male hormone, testosterone, and the bird resumes its song. In some tropical birds that sing (maintaining territory all year long), the testes of the males remain rather larger, presumably with high testosterone production throughout the year.

It has long been believed that birdsong is a form of advertisement. In the context of territory, it is meant to repel other males; in the context of reproduction, the purpose appears to be that of attracting eligible females. Again, it is likely that the song varies somewhat in its precise functions from species to species.

While numerous experiments have established the role of song in territorial defense, it is more difficult to determine its sexual function. Sexual and territorial functions are, in fact, probably interrelated, and in many species the structure of a song may incorporate these dual functions.

What we do know is this: In many species, song production greatly decreases (the gonads decrease in size and testosterone levels fall) following breeding, occasionally even ceasing altogether, which would help substantiate the idea that for some species, singing is a way of attracting willing females. Once the mate has been won, the incentive to sing isn't as strong. In studies of the male Prairie Warbler, experimenters noted that a male sang 2,304 times a day prior to the arrival of its mate, 965 times per day during the building of the nest, 192 times on the last day of its nest-building, 497 times per day during the laying period, 1,052 times per day during the incubation of the eggs, and 640 times per day during the nestling period. The Prairie Warbler is a long-range migrant, so it is particularly important for the males to quickly establish and then maintain a territory when they return in the spring.

In some species the extent of a male's song repertoire is an influencing factor in mate selection, with those males that sing the most complicated songs acquiring mates sooner than their less vocally diverse neighbors.

The Sedge Warbler is a bird whose singing ends after it finds a mate. During courtship, however, the Sedge Warbler's song is unusually long and complicated, with no two songs ever alike. In field experiments, the male warblers were recorded to determine whether repertoire influenced courtship success. Indeed, the males who sang the most complicated songs had an easier time attracting a mate than did those who sang simpler selections.

Not only does song seem to enhance a male's attractiveness to a female, but in some species it seems to stimulate the female's reproductive functions. Female Canaries whose males had large song repertoires built their nests more quickly and laid more eggs than females exposed to smaller song repertoires. And the female Ring Dove actually needs to hear the cooing of her mate if she is to produce eggs.

LEARNING TO SING

A bird does not emerge from the egg able to sing songs any more than a child is born speaking sentences. For both, the process is gradual and amazingly similar.

The young bird, like the child, begins to babble a subsong of meaningless syllables. Later, these syllables progress to become fragments – often mispronounced – known as *plastic song*. Finally, after much practice, the birds are able to properly articulate the songs of their species, or what is known as *crystallized song*.

The age at which a bird sings varies. Most young birds have frequently used begging calls. Some of these change as the nestling grows and fledges, so that songs or calls of adults may be based upon these early notes. Young Song Sparrows may start to sing as early as their fourteenth day of life. The young of many other passerine birds may begin within a month after hatching, but such early singing is not common because the young vocalist will come into conflict with its father.

Is the ability to make exquisite music genetic or learned? It appears that for most species – although not all – both influences are important. Like the young child learning to talk, most nestling birds learn language or song from adult birds.

Not long ago song development was thought to be simply inherited. Then, in a pioneering study of the Chaffinch, it was discovered that nestlings raised in isolation, who had never heard their species' song, did not achieve the song proficiency of those who had had repeated exposure to the melodies of their kind. The simple isolated song these birds produced had, in fact, little in common with the complex structure of the normal adult melody.

As part of the same experiment, the isolated Chaffinches were exposed to tapes of Chaffinch songs. If exposure occurred during the bird's first thirteen months of life, it was able to learn the songs. In cases where the bird was older when exposure first took place, the experimenter concluded that the bird had passed the critical period essential for learning song – a period that varies among species. In some species such as the Swamp Sparrow, there is a critical, short learning period at the start of breeding each year. Consequently, older males have a more varied repertoire than younger birds.

As in deaf humans, the degree to which a deafened bird can sing depends on how much song it memorized prior to becoming deaf. A nestling Chaffinch, for example, deafened when it is three months old, will learn to sing, but the quality and complexity of the hearing bird's song will be lost. The closer the bird is to adulthood at the time of its hearing loss, the better the chance that its songs will be normal.

There are a few species, however, for which song seems to be completely innate. A Domestic Chicken, Domestic Turkey, or Ring Dove deafened shortly after it hatches will still develop more or less normal vocalizations.

Any human baby born anywhere in the world could be transplanted to the other side of the globe and still readily learn the language of its adopted homeland. But the degree to which a bird can learn another's language varies among species.

Some birds have been known to adopt the songs of foster parents. The Zebra Finch is one such bird. Raised by a male of a similar species, a White-backed Finch, the Zebra Finch developed the song of his foster father. After developing the song, the Zebra Finch was placed in a cage with other Zebra Finches, all singing their native songs. Even then, the White-backed song was so deeply ingrained that the Zebra Finch persisted in singing it.

In Nebraska, I once was surprised to hear a simple trill song emanating from a bushy pasture. The singer proved to be a Field Sparrow, which sings a very different, more complex song. Perhaps somehow this sparrow early in life "fixed" upon the song of a nearby Chipping Sparrow at the edge of the field.

While some species do learn songs of other closely related species, when given a choice between the songs of their own kind and those of another, most birds will choose the former.

To see how selective birds are in their song learning, experimenters tutored young Swamp Sparrows with a series of artificially constructed songs made up of Swamp Sparrow-like patterns with their fairly regular slow trill and the more complex Song Sparrow patterns, alternating between fast and slow sequences. The males were tutored twice a day between the twentieth and fiftieth days of life, the critical period for song learning in this species. Months later, when the adult birds blossomed into song, they had learned only the Swamp Sparrow sequences, attesting to an inherited basis for selective learning.

Although many species learn songs only during an early sensitive period, some birds are capable of expanding their repertoires well into adulthood. Called *open-ended learners,* these bird species seem to place importance on being able to acquire new songs from new neighbors, skills that equate with better breeding success or advantages in the defense of territory.

Canaries are open-ended learners. A Canary is capable of picking and choosing its song, adding new sounds here, discarding an old sound there. This species' repertoire increases at least into the bird's second year of life. A complex song usually means the singer is a more mature bird, a quality that, again, appears to attract Canary females.

DUETTING

D'Arnaud's Barbets, a species found in Africa, have mastered the art of singing in harmony.

The birds live in groups of three to six, two of which are the dominant male and female. During the breeding season and to some degree throughout the year, the male and female sing an intricately synchronized song known as duetting, a form of vocal interplay that may be used to defend territory, cement the pair bond, and even help stimulate and synchronize the male and female reproductive cycles.

The male sings his part, then the female hers, and so on for several minutes until the song is finished. Other group members may chirp in with a few notes here and there, but overall they do not sing.

Something interesting happens, though, if the male of the pair is removed after the song. One of the other males wastes no time in moving in. The bird immediately begins singing a rendition of the song that up until that moment was reserved for the dominant bird. The female sings her part and before the last notes have been sung, the two have begun to form a new pair, as they begin to synchronize their duet more precisely. It also was possible, after the male was removed, for a human using a tape recorder to play just the male notes; the dominant female began attempting to match her notes to the tape recorder and in this manner formed a duet. But if the former male is returned, the new bird is driven back to being a subordinate member of the group once more.

Sometimes a secondary male and female may duet when the primary couple's backs are turned. Or they may go off to a distant corner of the territory to sing their song. In some cases, the primary pair may allow the two other birds their song. But since singing is an indication of dominance and territoriality in this species, the competition often is not tolerated and the primary couple will attempt to push the other birds out of the territory. In some instances, the secondary pair are able to carve out a piece of the territory as their own.

SONG DIALECTS

In the musical "My Fair Lady," Professor Henry Higgins, who himself spoke the Queen's English with the precision befitting his university training, attempted to rid Eliza Doolittle of her Cockney accent. If an equivalent exists in the world of bird studies, it is undoubtedly those ornithologists who have experimented with various song dialects in birds, attempting to determine how and why a bird in one area sings with a different "accent" than one across town.

As the Kentucky miner has his mountain drawl and the Brooklynite his nasal twang, so also does the White-crowned

Sparrow of Berkeley, California, have his distinctive whistling song pattern, different from that of his compatriot at nearby Sunset Beach. Like humans, birds of the same species sometimes have local dialects, making one population's songs different from another, even one that is geographically close by.

While song dialects have been found to occur in many species, the Bay Area's White-crowned Sparrows' dialect is probably the most examined. Sonographic studies of the sparrows' songs have shown marked differences between the song patterns of birds who live in three different parts of the San Francisco metropolitan area.

Are such differences in dialect detectable only with the fine-tuned aid of sophisticated scientific equipment, or can a bird tell the difference?

The Great Tit of Afghanistan sings an entirely different song from the Great Tit who lives in Germany. When recordings of the former's songs were played to the German bird, the bird might as well have been listening to a chicken cluck. So, yes, it does appear that birds do indeed notice a difference.

The obvious question is how these song differences occur, particularly in situations where the distance is not great. In an experiment to answer that question, one group of wild young birds (under one hundred days old) and one of nestlings were raised totally isolated from sound. The nestlings grew up to produce abnormal sounds, while the older birds who had had previous exposure to their dialect eventually produced sounds similar to their species' song, suggesting that dialect is not inherited but learned. The working hypothesis, then, is that during critical song-learning periods, some species are adaptable to learning even "foreign" dialects. While many birds will reject the songs of even a closely related species, they are capable of learning variations of their own songs.

While the fact that song dialects exist has been well documented, their function – if, indeed, there is one – is less clear.

Some ornithologists believe song dialects serve no real purpose. Others suggest that dialects may serve to keep the population "pure." Birds learn the dialect of their area and then settle within that area instead of traveling farther away where the dialect may not be known. A female chooses her mate by matching his song with the one committed to her memory.

In the well-studied paradigm, the White-crowned Sparrow, there is some evidence that males and females show a stronger response to their own dialect. Moreover, analysis has shown genetic differences between various populations of White-crowned Sparrows with different dialects, which suggests limited breeding between populations.

However, in a study of both sedentary and migratory populations of color-marked White-crowned Sparrows, females mated with males who produced a different dialect. Normally, female sparrows do not sing unless induced with doses of the male hormone testosterone. To see if the females would sing in the same "language" as their mates, they were injected with testosterone. Only a few of these females did, in fact, sing the same songs as their mates.

At the same time, male California birds often sang dialects from up and down the entire West Coast, yet mated and successfully bred with California females. Thus, in this species at least, song dialect did not seem to influence mate selection to the degree to which it was suspected. Yet it does suggest that females in a local population might select males that happen to use a somewhat variant song and that, in fact, dialects are continually undergoing change.

Nursery-school children play a game in which, sitting in a circle, one person whispers something to the person next to him or her, and that person is supposed to repeat the sentence to the next, and so on, until the message has made its way around the circle. As anyone who has ever played this game can attest, the original message is rarely conveyed to the last person. Song dialects evolve in much the same way.

Unlike young children, the Chaffinch normally copies its songs verbatim from other birds or tape recordings. However, sometimes an element is missed or one is added on. Such a change is passed on to other birds in the population. Another error, another change. In a decade or two, the songs of an entire bird population may bear little or no resemblance to the songs of their forefathers.

MIMICRY

A Superb Lyrebird lived for almost twenty years on an Australian farm. During the course of its residence there, this bird learned to imitate barnyard sounds. Almost nothing escaped its mimicry: the squeal of a pig being butchered, the sound of a chainsaw eating through a tree, dogs howling for their dinner, the farmer's wife playing the piano, the clip-clop of a horse's hooves. The bird's mimicry prowess didn't stop at mere sound effects. From time to time this avian imitator was heard to call, "Look out, Jack," and "Gee up, Bess."

Although overall the number of species that mimic is not very great, it includes birds from very diverse families – warblers, starlings, mockingbirds, lyrebirds, thrushes, bowerbirds, and others. Within some families, whole groups seem to use mimicry.

Most of these birds' mimicry involves imitating other species, although other sounds and even human voices have been heard. The average male Marsh Warbler mimics seventy-six species. In fact, sonographic studies of this bird's songs have shown that almost its entire repertoire has been pilfered from other birds. A Belgian study found that the warbler sang the songs of nearly one hundred European species as well as more than one hundred African species from localities where the bird spends its winters (and where it does not sing).

This doesn't mean that mimics do not have their own songs to sing as well. The total song repertoire of the mockingbird is made up of from 5 to 18 percent mimicry, the rest original; starlings spend about 10 percent of their singing time in mimicry; and approximately 53 percent of the White-eyed Vireo's songs reproduce the songs of other birds.

Why do some birds mimic? There are a number of theories, all of which may be correct because mimicry, like song itself, appears to serve a variety of functions in different species.

Because it appears that a large song repertoire may be helpful in attracting females, it makes sense that mimicking the songs a bird hears may be one way of increasing his repertoire, thus making him more desirable to females.

Another possible function of mimicry may be sovereignty over territory. One would expect that a good mimic such as the Mockingbird could ward off other birds by singing their songs. This did indeed occur when one mockingbird mimicked the alarm note of the Smooth-billed Ani, which caused the latter bird to leave the singer's territory as quickly as it could. Other studies, however, found that while mockingbirds with large repertoires of other species' songs were successful at keeping males of their own kind away, the mimicry had little effect on males of those other species.

Finally, in some situations it appears that mimicry may be useful as a protection device against predators. The female Burrowing Owl and her young, in situations where extreme measures are called for, has been known to imitate the rattle of a prairie rattlesnake, a sound that rarely fails to startle a potential predator.

Then there is the Thick-billed Euphonia, a small, brilliantly colored tanager. If the euphonia sees a predator near its breeding site, it mimics the mobbing calls of other species that are nesting nearby. As these birds answer the call and begin the task of mobbing the unwelcome intruder, the euphonia, like a general watching the battle from a far-off hillside, retreats to a safe distance and ventures forth again when the other birds have caused the predator to leave the area.

The songs birds sing – whether their own or borrowed or adapted – often serve to entertain and delight us humans. But for the birds themselves, a vast range of important functions are enhanced or accomplished, thanks to those trills and tweets and twitterings.

*A European Great Tit in a threat display to
another male. Note that the feathers are somewhat erected, the wings out, the tail spread,
and the bill facing its antagonist.*

6

The Quest for Territory

A male English Robin, the owner of an apparently fine territory, was put on alert that his area was about to be under siege. His warning was the melodic song of a rival male that had perched high in a tree right in the middle of the first robin's turf.

The newcomer sang a full, apparently equally fine song and then paused, during which the owner of the territory burst forth into his sustained vocal effort. The intruder then flew to another tree, with his unwilling host in pursuit. Atop a new perch, the newcomer once again launched into song, loud and long. Although he was in a position to easily catch the interloper, the owner instead replied with another song.

So it was that the two robins continued their confrontation throughout the territory. Two days later, the owner simply stopped the chase. His previously loud song grew softer and softer; finally, it ceased altogether. The bird flew off, abandoning what had been his territory, and the new owner took possession.

The eviction, like many, was accomplished without prolonged battles or bloodshed. While the owner of the territory occasionally delivered a few pecks during the chase, the new bird never touched his host, relying instead on persistence to wear down his adversary's resistance.

Not all evictions are this peaceable. On a similar April morning in the same forest, two other male robins battled in quite a different way. An unmated male invaded the property of his already mated nextdoor neighbor. Again, the prelude to the intruder's intentions was a burst of song, the loudest he could produce. After a few rounds of song, however, the battle turned ugly, with the uninvited bird furiously pecking the other. The two birds grappled, kicking with their claws and pecking with their beaks. Twice they lost their balance and fell to the ground, where, stunned though they were, the fight continued.

After a time, the defender of the territory seemed to be slowing down. Clearly, this bird was getting the worst of it. Most of his chin feathers had been pulled off, indeed, half his face was nearly bare. His opponent, on the other hand, looked as though he could go another nine rounds.

Two hours after the initial strike, the ailing bird simply stopped protesting. Instead of resisting, he fled from the attacker to sit on a high branch, singing quietly, while the land's new lord sang his loudest, flying from one tree to the next, establishing the limits of his kingdom.

The hen, the mate of the territory's former owner, had remained aloof during the war, but afterward lost no time in making her presence felt. She quickly began following the victorious bird; later, she mated with him and reared a brood of chicks.

As for the losing male, he hung around his former home for a few days, singing softly, occasionally being chased by the other bird, and looking dejected. After a while, he flew off and was never seen again.

❦

At some point, most birds establish some sort of "territory" or space that they lay claim to and will defend. Unlike the birds in the previous example, 95 percent of the time the owner of a territory is the winner of any dispute over territorial rights.

The concept of territory is nothing new. As early as 350 B.C., Aristotle noted that a pair of eagles needed a large territory and would allow no other eagle to encroach upon it. The English naturalist Gilbert White wrote in 1772, "During the amorous season such a jealousy prevails amongst male birds that they can hardly bear to be together in the same hedge or field."

As the study of birds has developed, so has the definition of territory. Simply put, a territory is any area a bird will defend. The size depends upon the species and can range from a sparrow's nest hole to several miles of hunting ground held by an eagle. The time a territory is under ownership also varies widely, ranging from minutes to a lifetime. And there are many different kinds of territories.

TYPES OF TERRITORY

Territories serve many purposes in the avian world. For some birds such as songbirds, woodpeckers, and birds of prey, the territory is used for courtship, nesting, and feeding. Seabirds, swifts, and swallows, on the other hand, use the territory for courtship and nesting, but feed in undefended areas. Then there are birds like the oystercatchers and bee-eaters that have separate territories for courtship/nesting and feeding.

Thus, a territory may represent different things to different birds.

The following is a brief description of some types of territories held by birds.

Mating, Nesting, and Feeding Territories. Probably the most common territory is the one used for mating, nesting, and feeding. In this multipurpose territory the male and female court, mate, and build the nest. The eggs are incubated here, and after the young are hatched, their food comes from the territory. Many bird species, including many woodpeckers, shrikes, thrushes, warblers, and sparrows, use their territories in such a broad manner and resident species may hold these throughout the year.

Mating and Nesting Territories. The Asian Scarlet Finch is a bird that uses a territory for mating and nesting but feeds with other birds on neutral ground. Grebes, a few swans, harriers, and Red-winged Blackbirds also will ardently defend their mating and nesting territories, yet travel elsewhere in search of food.

The Great Tit (right) prepares to attack a smaller Blue Tit (left), which is in a submissive posture.

Courtship and Mating Territories. Male Greater Prairie Chickens share a communal territory in which each male claims and defends a small area where he struts, dances, and generally tries to impress onlooking females in the hope that they will mate with him. The dominant male prairie chickens earn the best sites in which to perform their ritual. Other species that defend territories used strictly to attract mates include some birds-of-paradise, bowerbirds, manakins, some hummingbirds, the Sharp-tailed Grouse, and the Capercaillie.

Territories That Barely Extend Beyond the Immediate Nest. Many birds that nest in colonies of their own kind establish their individual nests and its environs as their territory and will defend it as such. Often the boundaries of the territory are dictated by the distance over which the bird's beak is capable of jabbing an intruder. Thus, the nests of the Peruvian Brown Pelican are spaced one per square meter, while the Peruvian Guanay Cormorants' nests are three per square meter.

Separate Feeding Territories. The Phainopepla is one of the few birds that has a feeding territory separate from its nesting ground. In southern California, this species primarily eats the berry of a plant that grows on chaparral-covered mountain hillsides, yet the trees in which it nests are found only on canyon bottoms. So the birds defend small courtship and nesting territories in the trees and fly into the hills every day to forage in their larger feeding territories.

Winter Territories. The Red-headed Woodpecker in Maryland was found to have a winter territory in which it hides large stores of acorns in tree cavities. The woodpecker takes this territory extremely seriously from September until May, and will defend its food bounty against potential theft by rival woodpeckers or other species of birds. After early May, however, the bird abandons its winter territory and moves to its mating and nesting ground.

Other birds who defend winter territories include some species of Arctic-breeding shorebirds on their Southern Hemisphere wintering grounds. Female adult Snowy Owls, wintering

Displays of the Common Black-headed Gull during courtship. At left, a gull shows aggression toward rival males (and at first toward females, too) with forward postures. The female stretches her neck upward to reveal her sex; the male becomes less hostile, and the pair move on to forward displays (bottom, left) and facing away displays (bottom, right). Thus the bills, which are possible weapons, and the black heads are partially obscured, reducing mutual aggression.

in Alberta, Canada, defend small territories rich with rodents, while the younger females have larger territories that are less abundantly endowed. Even some songbirds such as the Rock Thrush maintain territories in their subtropical or tropical winter areas.

Roosting Territories. Although roosting territories are the least studied of bird territories, it appears that a few species do perch at night in the same spot. Starlings appear to sleep in the same space every night, while in one experiment a European Treecreeper, which nests in a hole in the bark of a tree, attacked a stuffed bird placed at the entrance to its hole. Most woodpeckers occasionally visit the tree that houses their roosting cavity during the day and vigorously chase off other woodpeckers intent on entering their hole.

Group Territory. Sometimes a group of birds will work together to defend a territory. Groups of White-fronted Bee-eaters may defend several miles of foraging territory, while large colonies of Australian Noisy Miners are not above killing a bird who intrudes upon their territory. Most birds that have group territories will join in defense of that territory. Usually, such birds are of the same species.

Mobile Territories. South American Antbirds will follow a train of army ants as it marches across the land, dining on the insects flushed out by the long train during its overland trek. Naturally, the most dominant birds occupy the best and most prolific spots at the head of the moving column of ants, where insects frantically try to move away from the killer ants. The subordinate birds are left with the dregs, even though the ants may march through several adjacent territories, with shifts in attendant birds.

Another form of a mobile territory occurs in some species such as the rosy finch and Brown-headed Cowbird, in which the males follow a female and attempt to defend the space around her from other males.

Superterritories. Sometimes a bird will defend a territory that contains more food than it and its progeny could ever eat.

One might speculate that the rich territory acts as a sort of insurance policy against a harsh winter, ensuring that the territorial male is stronger and fitter than his potential adversaries. Under normal circumstances, bird population densities are such that a male is hard put to defend any more than the area that supplies basic needs – too big a territory becomes impossible to defend.

Yet most birds will try to defend as large an area as they can. That this depends upon the pressure of adjacent territorial birds is shown by behavior of the African Mustached Green Tinkerbird. In closed forests in Malawi, these birds maintain territories eight or ten times larger than they do in small, isolated forest patches.

THE SHAPE AND SIZE OF A TERRITORY

The Least Flycatcher has a territory of roughly seven hundred square meters. In contrast, the Golden Eagle's domain is some 93,000 square meters, 133,000 times the size of the smaller bird's.

These two birds epitomize the vast diversity found in territory size. Some birds, like the eagle, require a sweeping space in which to hunt for prey day after day; others, especially plant- or insect-eating birds, are able to do well in less extensive landscapes.

The shape of a territory also varies, according to the surroundings. Ideally, bird territories would probably be circular or hexagonal, with one bird's space abutting another's. But often natural boundaries dictate the shape of a bird's territory. If it abuts a river, ploughed field, or another boundary that does not require defense, the territory can be larger than one whose boundaries need to be defended. A bird whose territory is long and narrow has more boundary to be defended than one whose territory is of equal size but circular in shape.

Numerous factors determine the territorial dimensions needed by a particular bird. The function of the territory, whether it is used for food, the availability of other suitable habitats, and the time of year all are considerations.

Probably the foremost factor in determining territorial size is the bird's own size. Generally, a larger bird requires more space than a smaller one. Predatory birds, in particular, need more space if they are to feed adequately. Per unit of territory, the total mass of food decreases as a predator bird's weight increases, which means that as the bird's weight increases, so must the size of its feeding territory.

Two birds that are closely related are the Woodchat Shrike and the Great Gray Shrike. Yet the Woodchat Shrike's diet consists mainly of insects, while the Great Gray dines on small birds and rodents. Compare the size of their territories. The Woodchat Shrike requires between 10 and 30 acres to survive, the Great Gray between 115 and 245 acres.

The amount of food in an area is also a key factor in determining how much land a bird needs. If food is plentiful, the territory can be smaller than if the supply is scanty. The territories of hummingbirds vary markedly in size yet, flower for flower, most contain about the same amount of food. (Some hummingbirds and sunbirds defend optimal flower patches against their own and related species.) In an experiment to determine the influence of food on territorial size, half the flowers in a Rufous Hummingbird's territory were covered by bags. Not a day passed before the hummingbird had enlarged its territory to include just as many flowers as it had previously defended. When the original flowers were uncovered, the bird's territory shrank back to its original size.

Other elements of nature greatly influence territory size. In 1953, lemmings, which are the principal food of Pomarine Jaegers, ran rampant around Point Barrow, Alaska. The jaegers that year were content to defend territories ranging from 15 to 22 acres. The previous year, when their prey had been in short supply, the average jaeger territory was 112 acres.

Territorial size can also depend on competition for available space. When the density of the population is low, American Tree Sparrows use only 15 to 18 percent of their large territories. Most of their time is spent in the core section of the territory, which is surrounded by a defended but infrequently used buffer zone. In years of high population and increased competition for available space, the birds pack into smaller territories and the buffer zones are eliminated.

Other birds are equally adaptable. Many early-arriving male Little Ringed Plovers, House Wrens, and other species claim large territories that gradually shrink in size as later-arriving males enter the area. From the time of the first male's return until eggs are being laid, the territories undergo frequent, sometimes major shifts before achieving stability.

THE ADVANTAGES OF A TERRITORY

If you are a bird, a territory to call your own can mean the difference between breeding or staying a bachelor, the difference between having enough food or going without. If birds have the avian equivalent of self-esteem, the difference might be characterized as that between feeling as though one is king of the hill or simply the object of every other bird's pecks.

"To the victor goes the spoils" holds true in the bird's world as well as the human's, and one of the spoils for a bird is a territory of its own. The proportion of males of a species that actually breed varies depending upon the quality and extent of suitable habitat. In most birds, there is a small to large population of wandering "floater" birds that can move into a territory that loses its male or female.

The Red Grouse, a bird that lives in the heather of the Scottish moors, illustrates the difference a territory can make. Every autumn, males of this species compete for territory – in essence it is a battle for life because the winners will have enough to eat, while large numbers of the losers will die of starvation or be caught by predators. Only those who possess territory will breed the following spring. A hen that joins a male on a good territory will be well nourished and lay quality eggs. Ultimately, it is these chicks who have a better chance of surviving.

Although the acquisition of territory necessitates the burden of defense, the advantages usually far outweigh the costs.

Isolation. This first advantage is important, particularly during courtship and breeding. A male with his own territory can be assured of having privacy when courting a female within his domain. Because male birds are naturally antagonistic toward one other, this isolation also promotes a relative peace, greatly reducing the need for aggression, which, in turn, allows the bird to turn its attention to foraging and other important endeavors.

Spacing Out of Birds. If too many birds of a species were to gather in a spot where the food was easily obtained and the habitat friendly, it wouldn't be long before the food would be gone, famine would strike, and other food sources might be wasted. Also, predators would begin to concentrate their attention on such a site. Thus, the quest for territory actually serves to disperse the bird population, thereby making better use of all the available resources.

Every bird needs a minimum amount of habitat in order to survive, although this itself is dependent upon the quality of that habitat. Territory is one way in which the population maintains stability.

In 1951, in a forty-acre forest in Maine, experimenters sought to determine what would happen if territories were stripped of their owners. In this forest, 154 birds of various species dominated territories in early June when the experiment began. Over a period of three weeks, the birds were removed consistently, a measure approved by conservation authorities. Immediately, males without territories – floaters – began to pour into the area to stake their claims. In an attempt to keep this reduced population level, authorities continued to kill excess birds. At the end of the three-week period, 528 adult birds had been killed – three and a half times the original population! And yet the birds defending territories were as numerous as they were at the start of the experiment, indicating that a large portion of a bird population consists of floaters that are seeking a chance to breed.

The African Yellow-billed Shrike impales food on thorns (see also color photograph of shrike).

Ownership of Food Stores. An important advantage of territory is a monopoly on the food found in that area. Birds that do not have a territory are forced to pilfer from established territories, a pursuit that can be dangerous, or are pushed into marginal habitats in which they have to forage continuously to avoid starvation.

Reproductive Success. In some species male birds with large territories are more successful in attracting females than those with more meager areas. Such a bird is the Savannah Sparrow. One study found that males with territories of more than 600 square meters were markedly more successful in breeding than sparrows with smaller territories. Only 11 percent of birds in the small territories had nests, compared with 56 to 89 percent of the former group. Whatever its size, the territory must be able to provide food readily, so that the territorial male has a maximum amount of time for singing and defending the territory.

In some species territory appears to be a psychological necessity if a bird is to breed successfully. In an experiment with robins in aviaries, scientists found that only the dominant robin – that is, the one who had established the aviary as its own territory – was able to copulate.

Better Familiarity with an Area. Experiments are lacking on the advantages to birds of area familiarity, but researchers have examined the effects on mice. In one experiment, two mice were tested – one that had lived in a room for several days and one that was a newcomer. An owl was then allowed into the room. In twenty confrontations, the owl was able to capture only two of the mice who were familiar with the room, but eleven of the transients.

It stands to reason that a bird that lives in a given area quickly becomes familiar with the territory, learning the best places to forage for food, the quickest escape routes, where the strongest nest materials are in good supply, and, no doubt, a number of other things a bird needs to know to survive. A female mating with a territorial male benefits from the male's familiarity with the territory, and, of course, their offspring do too.

Expansion of Species Boundaries. To establish their own territories, younger birds often are forced to venture beyond the habitats occupied by their parents. Frequently, these pioneers must endure less than optimum conditions, spotty food, poor climate, and new predators. Of course, many fail to survive the rigors of these marginal habitats. In dispersal from the area where they were raised, the pioneers occasionally reach a suitable habitat beyond the range of the species, and are able to thrive, providing the species with an evolutionary bonus.

Psychological Bolstering. The possession of a territory is enough to make even the most self-effacing bird puff out its belly and strut. Consider the Common Black-headed Gull chick. Away from the nest the chick is pecked by every adult it encounters. Those that survive their early weeks are the lucky minority. Yet within the confines of its nest, as it gets fatter and larger, it puffs out its chest, opens its wings, croaks a shrill protest, and charges any adult who dares come too close.

The psychological shot in the arm associated with territory is not restricted to these gulls but is pervasive in the bird world. A bird in its own territory is a quite different individual from one that strays into a neighbor's turf.

To determine whether territory would influence behavior, a male European Robin was trapped in a wire cage within its territory. A neighbor, probably noticing the lack of song from diverse points in the other bird's territory, took advantage of his predicament and hopped over for a little snack. Although hardly in a position to do much, the first bird postured violently and burst into song, at which the trespasser meekly hopped home. Later, the still-caged robin was moved from his own territory onto his neighbor's. A transformation occurred. The bird that moments before had been as meek as a mouse turned into a lion and the first robin's former bravado vanished as he tried to make himself scarce within the confines of his cage.

CLAIMING A TERRITORY

In a colony of Great Skuas, a territory – a precious commodity in this densely packed group – became vacant when the pair who lived there died. Within one day, a new male and female, a pair who previously had no territory, had installed themselves. The pair moved slowly and cautiously, venturing out from the safe center into the periphery until they were turned back by their neighbors' protests. Each time they would escape back to the safety of the center, venturing out the next time in another direction. It was in this way that the skuas learned the approximate boundaries of their new land. Within two days the pair knew the limits to which they could safely go and were behaving as though they had spent their entire lives on the territory.

If this pair had not claimed the territory, another would have. Or, in the unlikely event that no bird was in need of a territory, neighboring birds would have expanded their borders to enlarge their holdings. In the skua colony, the birds without a territory spend time in a sort of "club," – an area where they rest, bathe, and preen. Most of the birds present are not breeding because they have no territory. Once in a while in heavily populated colonies with few territories, pairs form at the club and eggs are laid, but most of the time these usually younger birds form the reserve pool from which vacated territories are quickly resettled. If the colony occupies the entire area suitable for breeding territories, a few of the club birds may try nesting elsewhere and as a result new colonies sometimes are formed.

While most birds do not live in colonies like the skuas, it seems safe to suggest that most bird species have a large reserve of potential breeders that are constantly monitoring, waiting for their chance to take over should a territory become available. In the world of the Great Tit, any vacancy is filled by pairs who have been eking out an existence in the hedgerows outside the forest. And young European Nuthatches settle at the borders of poorly defended territories, waiting for the chance to move in.

Aside from the approach of being the quickest to act when a territorial owner dies or leaves, birds also gain territories by being the first to arrive in the spring. The first returnees usually are older, stronger males, who risk both unfavorable weather and poor food conditions in order to gain a prime territory. In migratory species the male usually returns first to the breeding area or may even stay there the entire year, with only the females and offspring migrating. The Eastern Kingbird is already paired when it arrives at its breeding ground, and both the male and female establish the territory. In some species such as the Pheasant-tailed Jacana of Asia, the Northern or Red-necked Phalarope, and various button-quails of the Old World tropics, the dominant female establishes and defends the territory, leaving the male to defend the nest, warm the eggs, and care for the young.

Wherever habitats are excellent, a Spotted Sandpiper female, who mates with more than one male, may claim a territory in which she "controls" several different males and thus several nests.

Some species establish group territories, with both related and unrelated birds working together to defend the land against outsiders. The Florida Scrub Jay often shares its territory with individuals other than its mate and offspring. The territory is owned by the monogamous pair but may have as many as six adult nonbreeders living on the property, all of whom help in defense. The territory is eventually passed on to an offspring or a sequence of replacement breeders.

Once a bird chooses a territory, it usually announces its ownership by singing or otherwise making its presence known. In the Arctic, the male Rock Ptarmigan makes display flights around his boundaries to let adversaries know that he is ready to defend his property against all comers. In establishing a territory, squabbles are bound to occur, particularly at borders. These gradually lessen as the borders become established.

FAITHFULNESS TO TERRITORY

If you're a bird-watcher (and even if you're not), you may have wondered at one time or another whether the birds you see in your backyard each spring are the same ones who were in residence last year.

There is no question that many birds become attached to their territories and if they migrate, they do return, at least to the general area.

With the use of numbered leg-bands, ornithologists have been able to trace the paths of many species to determine whether they return to their home ground. In one such study, 74 percent of migratory American Robins returned to within nine miles of their original homes.

This fidelity to familar turf is a common phenomenon in the avian world. When conditions permit, more than 60 percent of breeding male Song Sparrows and 12 percent of the fledglings return to the home locale in the spring. So the Song Sparrow that sings from a tree in your yard this year may not be the same one that occupied that spot the year before. But he may well have previously lived down the street or across town.

The female Pied Flycatcher's breeding sites may be seventy-five miles or more apart from year to year. If one brood is killed, she seldom returns to the same vicinity. Similarly, the average distance between the female White Stork's birthplace and the site of her first breeding is twice that for the male. Such dispersal is effective in preventing inbreeding – the mating of related individuals.

Then there are birds that are especially attached to their territory, returning to the same spot and even staying long past the time when the site is suitable. The Common Tern is one such bird. In a study of 2,964 terns on Cape Cod, 76.5 percent returned to the same nest site, the percentage increasing with each successive year. In one African forest, a bulbal was found at the same site at which it had been banded twenty years earlier!

An island that had once housed a huge colony of Arctic Terns eventually became so overgrown with shrubs that the terrain was unsuitable. Yet a few of the oldest terns refused to leave, nesting in conditions that normally would not have been tolerated.

DEFENSE

Birds are great trespassers. No matter how vigorously a bird defends its territory, other birds are continually encroaching, usually in the search for a few mouth-watering morsels. The owner of the land sings, displays, and in no uncertain terms makes clear to an intruder that it will be attacked if it persists. It may go away, yet minutes later, the same bird may be back, caught in the act once again. In so-called lek species, which include some grouse and the Ruff, the one or two most dominant males occupy the center of the lek or breeding arena. They fight frequently to retain their "territory" in the lek, for 90 percent of the females that come to breed choose these males.

It shouldn't surprise us then that with the privilege of territory comes the monumental responsibility of defense. As with much of life, benefits must be weighed against costs. Not that birds actually go through some kind of mental balancing, but the successful territory-owners that nature selects are always the strongest and fittest of the group. The Pied Wagtail seems to make just such a determination in approaching territorial decisions. From fall until early spring, the male wagtail defends a riverbank territory. Most of the bird's day is spent pacing the riverbank, back and forth, back and forth, picking up the insects that have been washed ashore. Because the territories tend to be large, typically the supply of insects is replenished by the time the wagtail passes by any given spot a second time.

Under normal circumstances, any intruding bird is driven away; to neglect to do this would greatly decrease the wagtail's chances of breeding successfully. However, sometimes the food supply is so plentiful that the wagtail ignores any intruder and goes about its business of eating. The chase simply is not worth the effort. Indeed, at times following a heavy rain in the tropics, the supply of winged termites is so plentiful that territories are temporarily suspended as birds from the area converge to gulp down as many of the nourishing insects as they can while the supply lasts.

The American Robin, like all altricial species, enters the world helpless. Naked, blind, and unable even to stand, young robins are nonetheless adept at begging for food. Without the constant nurturing of the parent birds, these robins would not survive. Don Enger/Animals Animals

A clutch of Domestic Chicken eggs in various stages of hatching. The chick in the foreground has almost succeeded in freeing itself, while the one to the left has just punctured the outer shell with its beak and now must use its body to break the egg. Embryonic birds develop an egg tooth, a sharp pointed tool on the tip of the upper beak, which assists them in puncturing the eggshell. This valuable tool falls off shortly after the bird is hatched. OSF/Animals Animals

◀ Overleaf: The Bee Hummingbird, at a weight of less than 1/10 ounce, is the smallest species in the avian world. Here, a male perches on a pencil eraser with room to spare. In contrast, the largest bird, the Ostrich, weighs approximately 330 pounds.
Robert A. Tyrrell, OSF/
Animals Animals

A parent bird will go to extreme lengths to protect its eggs or chicks from a predator. Here, in a common distraction display, an American Killdeer feigns a broken wing to lure an enemy away from its nest. Once the predator is a safe distance away from the nest and no longer a threat, the Killdeer will abandon its ruse and retreat to safety.
J.H. Robinson/Animals Animals

A nestling Brown Pelican plunges almost shoulder deep into the gullet of its parent to retrieve partially digested shrimp and fish the adult bird has caught. Many species feed their young in similar fashion. Through the use of its unique bill pouch, from which the young simply take out their meal, the parent pelican is able to carry more food to the nest in fewer trips. The adult bird's digestive enzymes aid in the chick's own digestion. Joe McDonald/Animals Animals

Every autumn an estimated five billion landbirds representing 187 species leave North America for the American tropics, while a similar number fly from Asia and Europe to the warm climate of Africa. Here, thousands of Snow Geese take to the skies for the annual migration.
J.H. Robinson/Animals Animals

Even bird species such as the Long-eared Owl that prey upon smaller birds and mammals may themselves assume intimidating defensive poses like this one. Predatory birds will carry small animals or other birds in their beaks or talons to the nest. The parent bird tears the meat into bite-sized pieces until its chicks are old enough to either eat the animal whole or tear it up themselves. Zig Leszczynski/
Animals Animals

◀ *The job of an incubating bird involves much more than simply sitting on its eggs for a certain number of days. The bird must be able to maintain a relatively even heat throughout the cool nights and warmer days. Here, an Australian Black Swan turns its eggs to ensure even heat distribution and prevent each egg's membrane from sticking to the shell.* Ken Cole/Animals Animals

These Western Grebe chicks not only get their dinner, but are carried to the meal on a parent's back. Some species carry their young in this fashion, particularly if a predator is nearby. Don Enger/Animals Animals

Emperor Penguins have adapted beautifully to brooding in frigid temperatures. To enable the young to survive the chilling Antarctic weather, the adults form a barricade of bodies around the young birds. The temperature at the center of this mass is as much as 20 degrees warmer than at the periphery. At the other end of the spectrum are the birds who live in tropical climates where the scorching sun is an omnipresent threat to young birds. Parent birds of many tropical species will use their wings as parasols to shield their young from the heat. Doug Allan, OSF/Animals Animals

The American Black-billed ▶ Cuckoo nestling has special mouth markings that arouse the feeding instincts of the parent birds. The young birds themselves have a strong gaping instinct for begging that ensures that their mouths are open and ready to receive any offering the parent may bring. Robert A. Lubeck/ Animals Animals

A Northern Gannet pair greet each other when one returns to the nest. Some birds who live in close quarters with many other individuals coming and going nearby call to their mates prior to landing so that they will not be mistaken for intruders or encroaching territorial neighbors, and attacked. Zig Leszczynski/ Animals Animals

At times in the animal world, small and large live together in remarkable harmony. In this case, Red-billed Oxpeckers occupy a comfortable perch atop an African (or Cape) Buffalo. Al Scabo/Animals Animals

This African Grosbeak Weaver, ▶ like virtually all the many species of weavers, is a master architect when it comes to nest-building. With nothing more than its bill and feet, the bird is able to knot, twine, and weave strips of grass to construct a complex and formidable fortress. Terry G. Murphy/Animals Animals

This Emperor Penguin chick hatched, only to find a warm home upon its parent's webbed feet. Securely tucked underneath a fold of belly skin, the penguin egg's internal temperature is maintained at 93 degrees Fahrenheit, despite the sub-zero weather in the Antarctic. Doug Allan/Animals Animals

The Ivory-Billed Woodpecker, ▶
long thought extinct, was spotted again in Cuba in 1985. Against all odds, a few of these birds survived then, but there is no guarantee that this elegant bird will continue to survive, given the transformation of its environment. None have been sighted since 1987. M. Stouffer/ Animals Animals

A Blue Tit indulges in a little holiday cheer, thanks to the local milkman. Certain individuals of some species of tits have learned to open milk bottles with their bills to get at the heavy cream that awaits just beneath the seal. This is but one of many examples of birds modifying their behavior to adapt to a changing environment, in essence, intelligence. John Paling/Animals Animals

A Common Cuckoo nestling, a parasitic bird that is hatched in another species' nest, ejects an egg from the nest of its foster parent, a European Dunnock. At hatching cuckoos have both the instinct and the leg, back, and upper wing muscles to lift and push any egg or hatching bird out of the nest. The end result is a nest empty of other eggs or chicks, thus guaranteeing the cuckoo a monopoly on the foster parents' time and energy.
J.A.L. Cooke, OSF/Animals Animals

A Gull helps itself to a Great White Pelican's egg. Birds must constantly guard against other birds who, given the opportunity, will steal an egg or even a chick from an unprotected nest.
Charles Paleck/Animals Animals

This pair of American Western Grebes perform a ritualized courtship dance on water, a necessary prelude to mating. Most bird courtship is highly ritualized and may involve intricate dances, aerial displays, posturing, and calls. Don Enger/ Animals Animals

The Great Gray Bowerbird got its name from its remarkable architectural feats in building bowers such as the one pictured here. Unlike most birds, the male bowerbird does not construct a nest in which to raise its young; that is strictly the domain of the female. Rather, the bowerbird constructs an intricately woven bower of sticks and vegetation, and then decorates it with colored stones, flowers, and the juice of berries all for the sake of attracting a potential mate. C.M. Perrins, OSF/Animals Animals

Although tool use in the bird world is rare, at least 30 species have been known to manipulate inanimate objects that enable them to better perform a task. This Great Grey Shrike enjoys a meal it has impaled upon a thorn. Roland Mayer, OSF/Animals Animals

The Great Frigatebird male puts on an extraordinary display, inflating his scarlet throat pouch. The female by his side has no red coloring on her throat.
Patti Murray/Animals Animals

A young Great Blue Heron practices flying. In some species, nestlings stand on the edge of the nest and practice flapping their wings, exercising their muscles for some time before they attempt flight. For some species such as the Turkey Vulture, these practice sessions are a necessary prelude to flying. In an experiment with Turkey Vultures confined to cages too small to allow any wing movement, ornithologists found that even after the vultures were released, the three-month-old birds were unable to fly.
Tom Edwards/Animals Animals

In times of severe shortage, the story is very different. Then the wagtail deserts the territory, gathering instead in flocks to forage. Even then, however, it is not inclined to relinquish its claim totally. The bird will periodically return to the territory to check the insect supply and to prevent another bird from moving in. Sometimes a compromise is struck, with the owner allowing another wagtail to forage, in exchange for its own defense of the territory.

Most birds are tolerant of other species, and in some forests as many as six species may be found nesting within several yards of each other. When it comes to their own kind, however, this tolerance usually ends. There is an old saying: One bush does not shelter two robins. For the male robin, the greatest enemy is not a bird predator, man's progress, or even the neighborhood cat. It is another male robin, a creature that is in constant competition for the same food, territory, and female. On the other hand, cavity-nesting birds may find available nesting sites in such short supply that they nest unusually close to one another, as in a group of Hispaniolan Woodpeckers, in which nineteen pairs had nest holes in one tree.

By far the most common method birds use to defend territory is song. In one experiment, male Red-winged Blackbirds were made mute in order to determine song's role in territorial defense. After the birds could no longer sing, neighbors became bolder than ever. Not only did they trespass more frequently, but their visits became longer. Once the birds regained their voices,they quickly reestablished their boundaries.

Another defense that may be used alone or in addition to song is a threatening posture or flight. Two male Hazel Grouse will chase each other through the air, and then may run side by side on the ground for as long as forty-five minutes in a dispute over boundaries. Other birds such as the Great Tit, the Skylark, and the Field Sparrow have specific aerial display patterns that they assume in defense of territory. Songs and other vocal signals are in effect acoustical displays, and frequently they are

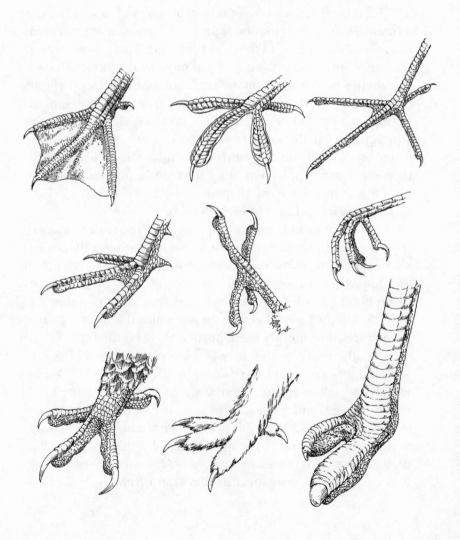

These are various bird feet, revealing adaptations for different functions, including swimming, walking, and climbing.

associated with particular visual displays such as the raising of the crest or the exposing of a normally concealed patch of color.

In some species a specific color is enough to incite aggression. Wave a red cape at a bull and he charges; show a robin a red breast (or any equivalent) and he postures wildly, puffing out his own breast as though it were a balloon. Many songbirds respond to mirrors or their image in a window or an automobile hubcap by pecking and attacking the "intruder." Even though it emits no song, the visual image of an obvious male may be enough for the territorial male to exhaust itself trying to drive away the competitor. One Variable Sunbird in Africa spent three to four hours a day pecking and trying to chase away its image in a window.

The Herring Gull that spies an unwelcome visitor on his turf adopts his most threatening posture and stance and slowly walks toward the other bird. When the trespasser refuses to budge, the owner bends down and rips out a large mouthful of grass, a challenge that the other bird may accept by doing the same. The birds may then sit opposite each other and attempt to pull the grass from each other's beaks. The battle can escalate until wings are beating against wings, beaks furiously pecking one another.

The vigor with which a bird defends its territory, and indeed the size of the territory, may depend on the time of year. Early in the breeding season a bird is much more likely to haul out the heavy artillery than when it is feeding young in the nest and later on, when it is molting. Many resident birds such as mockingbirds, flickers, and cardinals increase their singing as the young become independent and begin dispersing. This is the time when their populations are greatest.

In their fighting abilities, some birds seem fearless, to the point of taking on creatures several times their size. Usually these species are fast, adroit flyers. Some types of tyrant flycatchers will attack small mammals and even humans who enter their territory. The large European Capercaillie, a grouse, isn't reluctant to repel any human who dares approach its domain, as will male swans, and Sandhill Cranes have been seen driving caribou from their territory. Various terns, birds of prey, rollers, drongos, and Australian Magpies also behave very aggressively toward birds at or near their nests.

Physical combat in defense of a bird's territory is not common. In fact, various vocalizations and displays have evolved in order to reduce actual encounters. But when fighting does occur, it can be vicious. A penguin will defend its nest site from a neighbor by beating the intruder with its flippers and jabbing and biting with its sharp beak. At the end of one of these encounters, it is not unusual for both participants to be smeared with blood. Very rarely, two equally matched males will fight to the death over a territory.

One especially brutal fight was observed between two European Robins. The two birds fought until one actually killed the other, after which the victor continued to peck at the corpse. The observer examined the dead bird and saw that both its eyes had been pecked out and its skull exposed. The victorious bird perched high in a tree and sang its sweetest song for a while. Then it flew down to the corpse and began mutilating it again. Within a few minutes the bird had pecked a hole in its victim's side.

For birds, territory often represents life itself and sometimes is worth dying for, although ordinarily the conflict does not escalate to that point.

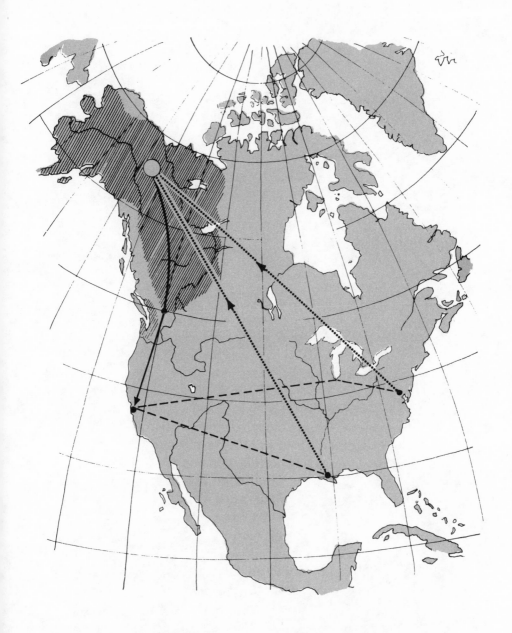

This map shows the path White-crowned Sparrows took to their normal wintering grounds
in San Jose, California, despite having been
flown by airplanes in previous years to Baton

Rouge, Louisiana and Laurel, Maryland.
The marked birds spent each intervening summer nesting in Alaska, having returned there
by the routes indicated here by the dotted lines.

7

Migration

On a seemingly ordinary autumn night, up to 12 million birds may pass over Cape Cod on an arduous, nonstop journey to their southern winter grounds.

Like marathon runners loading up on carbohydrates in preparation for a race, these birds have prepared their bodies for the abuse that several days without a break to feed or sleep will inflict upon them. Physically well-conditioned, the Blackpoll Warbler, but one species in this great mix of birds, takes to the skies only after it has spent ten to twenty days feeding, beefing up its weight from 1/3 ounce to 3/4 ounce. This is not to say that the birds plan their migratory program. Rather, it is genetically built into them, to the extent that hormones trigger appropriate behavior such as the heavy feeding at the appropriate time of the year.

Wave upon wave of the birds fly over Bermuda and then down to the Lesser Antilles, the first stage of the twelve-hundred mile journey. Reaching Florida, the warblers, perhaps with endurance waning, encounter strong trade winds from the northeast. This is just the help the birds need to enable them to complete the trip.

Eighty-six hours after leaving their breeding grounds in the northeastern United States, the birds that survive this test of endurance arrive with the wind at their backs in northern South America. Having neither rested nor eaten since the onset of their journey, they might be considered little more than living skeletons with feathers, their fat, protein, and body water having been depleted by the rigors of long-distance flight. Yet if they had to, these birds could travel farther (as some birds such as golden plovers do), for their bodies were sufficiently fattened to provide enough flying fuel for ninety-five nonstop hours. Nature has honed their systems for eons, and those birds who are weak and insufficiently fattened for migration drop to their deaths in the ocean.

Each year, these small birds and millions of others make a monumental effort, one perhaps unmatched by any other living creature. There is no known human equivalent for the rigors of bird migration. One must deal in hypotheticals: for a human being to match the metabolic exertion of the Blackpoll Warbler or any other bird who shares its migration, he or she would need to run a four-minute mile – for eighty hours straight!

The fall and early spring sky is a virtual freeway of avian activity. Although by no means all birds migrate, an estimated 5 billion landbirds representing about 187 species leave Europe and Asia for the warmth of Africa and tropical Asia every fall, while a similar number fly south from North America to the American tropics. In many species, males, females, and the young all migrate; in other species migration is selective, with more young than old leaving, more females than males, as young and female birds are less likely to survive a harsh winter, particularly given competition against older male birds. And in certain species such as the American Golden Plover, the adults leave on their migration southward days or even weeks before the inexperienced young birds begin the same journey.

Migration distance varies greatly. For some birds the distance is short and vertical. The North American Blue Grouse spends the winter in a mountain pine forest, descending on foot about one thousand feet in the spring to nest in deciduous woodland and feast on a lush crop of leaves and seeds. Similarly, the extent of many tropical hummingbirds' migration is a few thousand feet up or down a mountain. (A few hummingbirds of lower areas migrate upward after breeding to mountain grasslands that bear many flowers in July and August, then return to the lowlands before cool weather arrives.)

When humans think of migration, however, it is not these short treks that come to mind but the long-distance travel for which many species are renowned. Most species that migrate travel several hundred miles or more, often moving from one continent to the next. For a few, however, the distance of the full north-south-north migration may approximate that of the circumference of the earth.

If awards were given for distance alone, the Arctic Tern would surely rank highest. Every fall this bird, an inhabitant of the high latitudes of the Northern Hemisphere, travels eleven thousand miles, reaching the packed ice of the Antarctic. While this is no small accomplishment, the tern, being a seabird, is able to feed along the way, making the trip significantly less of an endurance test than many other birds' migration. In fact, the Arctic Tern reaches the Antarctic just about in time to leisurely commence the homeward trip north again, and so is almost constantly on the move.

Another bird whose endurance is put to the test every year is the Alaskan Bristle-thighed Curlew, which journeys six thousand miles across the Pacific Ocean to reach its winter home in the Polynesian Islands. For long stretches of its migration – the longest nonstop stretch being eighteen hundred miles – the curlew is unable to feed or rest because there is no land.

Over long stretches of oceans the migrants fly, across barren deserts, blown by hurricane and sandstorm, chased by predators. Finally, the lucky ones arrive. The others – sometimes more than half of those who started out (often the young, inexperienced birds) – never reach their winter grounds.

Still, every year they fly. How and why did this mysterious movement begin?

THE EVOLUTION OF MIGRATION

It is likely that birds have undertaken some form of migration for millions of years, certainly as long as there have been seasons. In equatorial regions, many birds regularly move into areas during and after the rainy season, then move out again.

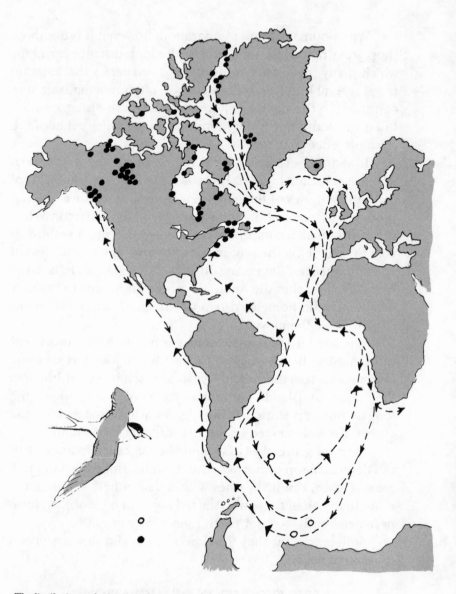

The distribution and migration of some Arctic Terns from North America. The most northern-nesting of these terns travel as many as 11,000 miles across the Atlantic and southward to Antarctica during their annual migration, and then another 11,000 miles back to their northern breeding areas.

Within 20 degrees of the equator in Africa, for example, millions of birds migrate, excluding those that migrate there from Europe and Asia. The Pleistocene ice ages probably accentuated migration as a means of adapting to seasonally cool versus cold climates. The almost continuous changes in the extent of glaciation most likely enforced migration over very long distances, with natural selection fashioning many well-adapted migratory species.

There are many theories regarding the origin of migration, many conflicting, some carrying little scientific weight.

One is that birds originated in northern latitudes where glaciation during the Pleistocene or even earlier forced them to fly south in search of more verdant fields. Conversely, some believe that birds originated not in the north but in the tropics and that increasing populations in some species and their demands upon the area's food supply, particularly during the breeding season, forced some species north every spring in search of suitable feeding grounds.

Another view held by some is that migratory pathways originated in the drift of continents northward from Antarctica, and that migrating birds today are in essence attempting to return to their ancestral homelands. The main problem with this theory is that the major continents drifted apart millions of years before most modern bird species evolved.

Although no one can say with any real certainty how and why migration evolved, it is safe to say that evolution has undoubtedly affected migration. Most species are long-time veterans of glacial movements.

To determine whether the tendency to migrate is genetic in some bird species, researchers performed a series of experiments, interchanging eggs between British colonies of the non-migrating Herring Gull and the migratory Lesser Black-backed Gull. Nine hundred young birds were reared by foster parents of the opposite species. The young birds were banded and later recovered by capturing them. Researchers found that many of the nonmigratory gulls raised by migratory foster parents followed their foster parents, migrating to France and Spain. Perhaps more surprising was the finding that the migratory nestlings reared by nonmigratory foster parents also migrated, even though their parents stayed in Britain.

Thus, for many species migration would appear to be a genetic trait.

Interestingly, in some species of birds there have evolved both migrating and nonmigrating populations, along with populations in which some individuals migrate while others stay in place. In southwestern Germany, one-fifth of the European Robin population lives the entire year within a couple of miles of its breeding territory. The birds exhibit no signs of migratory restlessness – a common trait among migrating birds; nor do they gain weight as fall approaches. In contrast, migrating robins in the same geographic area put on large stores of fat and show a restlessness known as *Zugunruhe* prior to embarking on their journey to winter grounds. Likewise, some Blue Jays in the New York region migrate south some hundreds of miles, while others of the same population stay in place throughout the winter.

Why is there a major difference between birds within the same species? Winters in southwestern Germany and in the northern United States run the gamut from mild to severe; mild winters favor year-round avian residence, while harsh ones make migration advantageous. Thus, the migratory habit evolved with physiological "triggering" in some birds but not in others. In very severe winters, however, almost all the robins and Blue Jays in these populations migrate.

Sometimes birds acquire or lose the migratory habit. The Canary-like Common Serin has moved north from its Mediterranean ancestral grounds in the past century. These northern Common Serins are migratory, unlike the serins that still live around the Mediterranean. On the other hand, several Northern Hemisphere birds such as the Common House-martin and White Stork long have spent the winter in southern Africa and then migrated north to breed. In recent years, however, a portion of these wintering birds have so adapted to parts of southern Africa that they breed there and have become resident instead of migratory.

ADVANTAGES OF MIGRATION

The Stonechat and Whinchat are two similar thrushes that have very different migratory habits. The contrast between them helps illustrate the delicate balance between braving the rigors of migration or staying behind to face cold weather and a scanty food supply.

These two birds dine on a similar diet of insects in the open grasslands, heaths, and recently planted forests of Britain. In the fall, the Stonechat either remains in Britain or, in some migrating populations, travels as far as southwestern Europe. The Whinchat, on the other hand, is a long-distance migrant that journeys south of the African Sahara Desert for its winter respite where it meets and overlaps with tropical African *resident* Stonechats.

An especially cold winter in Europe means that many Stonechats perish. Yet by remaining near their breeding grounds, surviving Stonechats are able to start nesting early, laying three clutches of eggs to the late-arriving Whinchats' two. By producing more young, the Stonechats are able to compensate for losses from a cold winter.

As for the Whinchats, undoubtedly many are lost in flight between the breeding and wintering grounds. However, the journey probably is no more hazardous than remaining in Britain to face winter's cruel challenges.

The bottom line is that, despite their differences, both species are thriving, thanks, in part, to an often mild British winter which ultimately serves to lessen the Stonechats' losses.

The whys of migrations continue to puzzle scientists. Clearly, in order for a bird to leave the familiarity of its breeding ground and embark on a risky journey from which it may never return, there must be both disadvantages to staying and advantages to leaving. Obviously, to insect-eating birds, the chance of there being any deep, long-lasting snow weighs heavily as an overwhelming disadvantage.

One factor may be that life in a tropical climate appears to be conducive to a bird's survival. A bird living in the tropics has an 80 or 90 percent chance of surviving a year. Those who must migrate to get there reduce their survival rate to about 50 percent, while the chances of survival for those who live the year in northern climates ranges from about 20 to 58 percent. This is reflected in the size of the clutch – high-risk, northern-summering species lay clutches averaging more eggs than do related species that are tropical residents.

In the past, migration has always been considered in terms of escape from the adversity associated with poor climate, lack of food, increased competition for food, and shortage of nest sites. Today, however, many ornithologists view migration in a different light. Instead of escape, migration may be a bird's way of actively exploiting – in modern parlance, "optimizing" – favorable opportunities in a different land.

The place where birds nest and raise their young has been thought of as their "home." Yet many species spend more months in their nonbreeding grounds. The Blue-winged Warbler, for example, spends only two or three months in North America, remaining the rest of the year on the wintering grounds in Central or South America, or the West Indies.

Should these birds, then, be regarded not as northern birds, as previously thought, but as tropical natives exploiting the abundance our summer has to offer? Many scientists believe this may be the case with some birds that breed in the Northern Hemisphere.

There are about four hundred species of Old World warblers, many of which live year-round in Africa; only about one-eighth of the birds make long-distance migrations. Among strictly New World groups, less than half of the 114 species of wood warblers (unrelated to Old World warblers) and only 30 out of 375 tyrant flycatchers breed in North America and migrate southward as far as South America, and of 390 hummingbird species, only 21 are breeding visitors or residents in the United States, and only 4 reach Canada.

These figures suggest that their nesting area is not the true home of these species and that they, in fact, are derived from tropical ancestors.

No matter the bird's home, one of the most important migratory advantages for many birds is that it allows them good year-round feeding opportunities. In many cases, the migrating bird abandons what promises to be a cold northern winter with little or no food for a moderate one in the south. Or birds of

By varying their forms of flight, birds can save considerable energy. In undulating flight (top), the bird descends in a glide, interspersed with brief bouts of wing flapping to regain height. Bounding flight (bottom illustration) involves flapping bursts that alternate with passive periods in which the wings are folded. The wing shape usually dictates the appropriate, normal flight in any species.

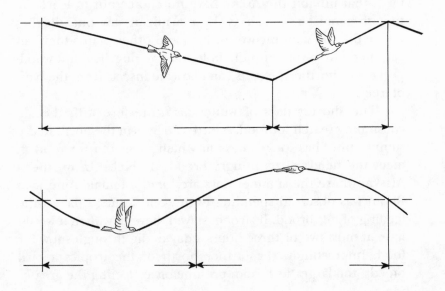

the southern part of the Southern Hemisphere may migrate north in the southern fall and south in the southern spring. In contrast, the migration for tropical birds from regions with seasonally varying rainfall may be immediately prior to and after the dry season.

Food is likely to be the main factor in determining migration. Winter is a death knell for insects. Thus, most insect-eating birds of the north must be early and long-distance migrants, the exceptions being some birds that can dig their way into the dens of hibernating insects. Some birds, however, may go south only if their food supply runs short, and only far enough south to find food. Many seed-eaters such as finches and predatory birds like the Snowy Owl and Rough-legged Hawk are what is called irruptive species, migrating only as need be. In these species, the main items of food (seeds, mice, lemmings) undergo cyclic shifts and may be abundant one winter and scarce the next.

Not only is there less available food in the winter for birds who live in northern climates, but it is more difficult to find. Frozen water and soil and a layer of snow all form a barrier between many birds and their food source. Geese and shorebirds that inhabit the Arctic have no choice but to leave, although some birds such as the ptarmigan stay put, digging under the snow to uncover willows and other plants they can eat, then burrowing into the holes they've dug. Tits and woodpeckers of northern forests can extricate insects from the bark of trees.

The shorter days of winter also interfere with the all-consuming search for food. A bird who lives in the north in the summertime has longer days in which to search for food to meet the needs of its hungry brood of chicks. In northern Alaska, where the summer days are long, a female American Robin was observed spending twenty-one hours a day in the feeding of her brood. If a robin stayed there in winter, it would have at most two or three hours a day to dig through snow for food. Interestingly, the farther north of the tropics a bird breeds, the larger its brood, presumably reflecting the greater risk and loss of birds during migration and on wintering grounds.

Another advantage of the geographical movement known as migration is that it often provides an opportunity for birds to eat a more varied and nutritious diet. It isn't far-fetched to hypothesize that some element in a bird's summer diet may be especially valuable to the young nestlings, or that the winter diet may contain nutrients that aid in reproduction.

If the migratory birds of the world were suddenly to "decide" to remain in their winter quarters, think of the demands on the food supply and territory. While millions of extra birds competed with resident birds for seeds and insects, and for territories, the bounty of the far north would be wasted. Thus, many species have evolved in such a way that they are able to live in two or more different habitats, exploiting the good from each, and reaping the benefits of both.

This is the way the Pennant-winged Nightjar, an eater of termites and other flying insects, uses migration to its advantage. The nightjar breeds during the rainy season from August to November in central-southern Africa. In February and March, these nightjars migrate to tropical Africa north of the equator, where the rainy season usually begins. By following the rains, the Pennant-winged Nightjar is placed in precisely the best position to enjoy its favorite food at its most abundant.

PREPARATIONS FOR MIGRATION

A few days before the tiny Ruby-throated Hummingbirds are ready to cross five to six hundred miles of open water in the Gulf of Mexico, both males and females embark on a campaign of binge-eating, sucking up every drop of nectar, every insect and spider they can get their beaks into. By the time each hummingbird is ready to beat its wings for long-distance flight, it has doubled its weight, expanding to a full six ounces.

The hummingbird is not alone in this gluttony. All birds that migrate must take on extra fat prior to setting off, a preparation sometimes referred to as *Zugdisposition*. The amount varies, according to whether the flight will be long and whether food will be available along the way. Diurnal migrants such as swallows and hawks feed as they migrate and so put on little fat prior to migration. The addition of so much extra fat may well handicap some birds in feeding, which could explain why night migrants cover so much distance so quickly.

The average amount of fat on a small nonmigrating bird is from 3 to 5 percent of body mass. A bird whose annual trek is several hundred miles (a short-range migrant) will boost its fat reserves to between 13 and 25 percent of body mass, while a bird that flies from one continent to another may take on so much fat that it accounts for almost half its weight.

Having seen what migration costs a bird, it is easy to understand why this massive build-up of fat is necessary. When scientists measured the body weight of more than two thousand small migratory birds belonging to nine species drawn to a Lake Ontario lighthouse during migration, they found that the average bird lost 0.91 percent of its body weight for every hour of migration.

The White-throated Sparrow increases its body fat to 17 percent of its weight prior to beginning its migration. By the time the bird is safely ensconced in its winter grounds, the rigors of its journey have reduced its fat content to 6 percent, a figure that eventually climbs to 12 percent during the course of the winter.

So birds eat, consuming enormous quantities of energy-rich food, some of them stuffing themselves to the point of obesity.

But what is it that tells a bird it is time to begin to prepare itself to leave for far-off lands?

It seems that some birds are predisposed to migrate, and that certain external factors stimulate this behavior. This predisposition appears to be a cyclic mechanism closely tied to hormonal levels and reproduction, which ensures that the bird leaves for its breeding grounds just proir to the time of year most favorable for raising its young.

Although theories about migration triggers abound, one can only speculate. However, it may be safe to say that for northern birds, hormones control the timing and sequence of events in the following manner. After a winter of short days and untapped reproductivity, the pituitary gland activates. The longer, warm days of spring (for birds wintering well north of the equator) make it easier for a bird to find food and reduce heat loss, two factors that mean more energy. Increased daylight also appears to boost the flow of two hormones, corticosterone and prolactin. Together, these hormones set in motion a whole set of migratory preparations, including the development of the bird's sex organs, an increase in fat, and the interesting phenomenon known as *Zugstimmung*.

If *Zugdisposition* is the physiological preparation the bird undergoes, *Zugstimmung* is the sum of the changes in behavior the bird must adopt to begin and sustain its long flight. In captive birds, *Zugstimmung* takes the form of *Zugunruhe* or nocturnal restlessness. In experiments with captive migrant birds, scientists recorded activity graphs of the captives, in which they slept for fifteen minutes to two hours after sunset, only to awaken and begin hopping and fluttering about the cage until midnight, after which their activity died down. This restlessness occurred at about the same time free migrants were flying.

In a California laboratory, an interesting aspect of *Zugunruhe* was tested when migratory White-crowned Sparrows, caged under the open sky, perched facing south at night during the fall but north in the spring when the species would be flying back.

Some studies have shown *Zugunruhe* to be related to the length of day. Captive Bramblings, throughout their normal spring migration, were kept in an environment that mimicked the short days of winter. The birds exhibited no restlessness nor did their gonads enlarge. It was not until their daily dose of light was increased to fourteen and a half hours that they began to prepare to migrate.

Zugunruhe, however, is more complicated than a bird's desire to migrate when the light is just right. In one study, two similar Arctic species, the Willow Warbler and the Chiffchaff, were hand-raised and later, as adults, placed in cages. The Willow Warbler spends three or four months en route to Africa, while the Chiffchaff's journey to southern Europe or northern Africa is accomplished in a month. Over the course of the study, both species were kept in cages with natural light and artificial light (twelve hours light, twelve hours dark). Regardless of the amount or source of light, all the birds developed night restlessness and began to gain weight at the same time and for the duration of their individual species' migration.

These former nestlings had no previous migratory experience, and yet something within them just knew when the time was right, as well as how much activity it would take to get the bird where it needed to go.

Fattened and psychologically prepared as the bird may be, the weather doesn't always cooperate. No matter how eager a bird is to begin its migration, most will delay the trip if bad weather is on the way. A logical question is How can a bird possibly forecast the weather? It's impossible to know for certain but birds appear to be very sensitive to changes in barometric pressure; some birds such as pigeons, are also sensitive to infrasound, which meteorologists frequently use to track the weather.

Regardless of how birds do it, they just do. In the spring, a major northward movement of birds traveling to their breeding grounds in the United States often will coincide with a lowering of barometric pressure toward the southwest, followed by a strong flow of warm southern winds from the Gulf of Mexico.

ORIENTATION AND NAVIGATION

Every year in what has to be one of the migratory wonders of the bird world, the young New Zealand Bronzed Cuckoo takes off, one month after its parents have left. The young bird flies more than one thousand miles west across the ocean to Australia, then changes its course northward for another nine hundred miles until it reaches the Solomon and Bismarck islands. There, it may come upon its parents, already settled in.

How can young birds without benefit of their elders' experience successfully navigate the black skies of night over seemingly featureless water? Many birds may, in fact, learn aspects of the migration route and destination from the elder birds. As for those like the New Zealand Bronzed Cuckoo, one can only suggest that a hereditary component must be involved, and no one yet knows for certain how these young birds accomplish such an amazing feat.

One of the most perplexing questions in ornithology is how any migrating bird can find its way home. Consider the elements working against them. Most birds migrate at night, when any help their sense of sight can provide is sorely limited. Even when their visibility isn't obscured, there may well be little to see but long stretches of water or sand.

Yet every year millions upon millions of birds travel hundreds, even thousands of miles, sometimes setting down among the same trees they frequented the previous year. Even more amazing are the homing abilities of birds that have purposely been displaced from their homes. Try to imagine what it would be like being blindfolded, driven for hours or even days in a car or airplane, and then released, somewhere in an open field in the middle of nowhere, without a clue where you are. How would you find your way home? A Manx Shearwater did, flying back to its home in Wales in the British Isles, in only twelve and a half days after being released in Boston, Massachusetts. So did marked White-crowned Sparrows that were shipped to Baton Rouge, Louisiana. The birds returned the following winter to their wintering grounds in San Jose, California. Once again, they were caught, this time shipped to Maryland, and once again they made their way back to California at the appropriate time for their usual return.

For a migrating bird to travel the skies it must, of necessity, "know" where it is. It must in some way fix upon the direction where its goal lies. It must be able to stay on course toward that goal, and it must fly at the altitude that enables it to use the least energy. Finally, it must know when to stop flying.

There are several ways in which a bird uses its own senses and nature's many cues to do this. Many birds seem to prefer one information source, relying on others only when the favored one is unavailable. Most migrants use at least two navigational modes, a duplication that reflects the great risk involved, and the danger of "putting all one's eggs in one basket."

Visual Landmarks. The young bird venturing from the safety of its nest spends a good part of its time wandering, getting a sense of the lie of the land. When the bird finally does set off on its migration, it has a fairly good idea of the landmarks it will use to help guide it home in the final stage of the return trip.

Landmarks are quickly learned in many species. In experiments with Homing Pigeons, scientists found that after only three to five releases, the pigeons were able to navigate the ten miles to their home loft. When the release point was between twenty-one and thirty-three miles, however, the birds could not seem to orient themselves properly. Yet when the distance was increased to more than forty-six miles, they had no trouble flying home.

The explanation for this paradoxical finding is that within close ranges, the pigeons "pilot" by landmarks, but for distances beyond forty-six miles, they must have an inborn map or grid so that they can effectively establish their coordinates and those of their destination. Armed with these built-in capabilities, they are able to determine the correct course to fly. For middle distances, it appears the bird has no such method of orientation.

In some species that migrate during the day, visual landmarks may be all that is necessary to guide the bird to its destination and back. The young goose learns from its elders the

traditional migratory route of its species. Once the bird somehow fixes in its brain the forests, shores, rivers, mountains,
prairies, and so on, it has all the information it needs to reach
its destination. Geese usually have traditional migratory stopping places, and family groups fly together, so the adults' store
of information is passed on. Nocturnal migrants, on the other
hand, rely primarily on other navigational cues. However, when
daylight comes, they, too, observe their surroundings for familiar landmarks.

Birds that fly in flocks also tend to visually orient themselves better than those setting off on solitary treks, presumably
because the birds pool their navigational skills.

*A male European Whinchat in the process of
landing. During landing, many birds spread
their tails and fan the air with their wings,
turning forward the large flight feathers on
their undersides, to slow themselves down.*

137

In addition to topographical information, many birds may use ecological and meteorological clues for orientation. In the northern temperate zone, spring advances northward about 2 degrees of latitude per week during the months of April and May. Thus, the difference one week's sunshine makes in the brilliance of flowering orchards and other trees may be discernible to a bird, a clue that it took a wrong turn that threw it off course.

As one can imagine, a bird migrating over thousands of miles of ocean has few visual landmarks, but even these birds are aware of visual cues. The way in which the wind is blowing, cloud formations, the shapes of islands and reefs, and whether or not the bird sees any marine life are all possible route markers for the migrant bird.

Navigation by the Sun. Landmarks are not the only signs the migrating bird uses to find its way. Scientists had long suspected that some birds may navigate by the sun, but proof of this skill has only come in recent years.

How does a bird do this? For a bird to use the sun as an effective compass, it must somehow "know" the time of day. The position of a given point on earth relative to the sun changes by 15 degrees each hour. To orient consistently in one direction, somehow the bird must understand the sun's position relative to direction as the day passes. Birds, like most living things, have an internal clock that regulates their daily cycle of activity and rest. It is this clock that enables the bird to use the sun as a directional guide.

In one experiment to determine whether birds do, in fact, compensate for the apparent motion of the sun, researchers trained starlings and some other species to feed from the northwest-positioned cup of a series of cups placed around the perimeter of a circular cage. When the birds could see the sun, they easily chose the correct cup. The birds were then trained to accept a stationary light bulb as a sun substitute. After that, they fed from cups progressively farther to their left, continuing to compensate for the progress of their surrogate sun as the day passed.

Homing pigeons, long bred specifically for their navigational abilities, also appear to use the sun to help guide them back to their lofts. Released under a variety of weather conditions, homing pigeons are able to fly home when they can see the sun but have fared poorly in experiments done on fully overcast days. Not only do these birds use the sun as a compass for determining direction, they are able to compensate for its changing position.

The Adélie Penguin, a flightless bird, also relies on a sun compass. When a group of penguins was taken from their coastal breeding areas into the interior of Antarctica and then released, the penguins wandered randomly on cloudy days. But when the sun came out, they headed north-northeast back toward their breeding grounds.

Celestial Navigation. That birds use the stars to navigate was first demonstrated in 1957 with hand-raised Garden Warblers. The birds were housed in circular cages in a planetarium. When it came time to migrate, the researchers watched the birds through the glass bottoms of their cages. The warblers faced north in the "spring" and south in the "fall" of the simulated night sky of the planetarium. When the fake stars were dimmed, the birds became disoriented. Later, the researchers rotated the planetarium sky 180 degrees, and the birds also reversed their internal compasses to compensate for the shift.

In another experiment aimed at identifying the stars Indigo Buntings use for orientation, various constellations were blocked from the birds' view. The experimenter assumed the birds used the North Star to guide them but found instead that they used constellations within 35 degrees of the North Star. Moreover, when one constellation was under cloud cover, they used another one, a testament to the birds' flexibility, a useful trait given the ever-changing night sky.

Geomagnetism. A bird often flies in dark skies with nary a star in sight and very little else to give a visual reading of its whereabouts. It isn't surprising that the bird flying through clouds, weaving in a twisted and tentative pattern through the blackness, does not have as accurate an orientation as it otherwise would. Yet radar studies of birds have shown that despite adverse conditions, they can maintain the basic correct course for hours.

That birds are sensitive to the earth's magnetic field is an idea that for many years met with much skepticism. Then in 1965, a team of ornithologists showed that the orientation of European Robins placed in solid steel cages that concealed the sun and stars could be altered by imposing an artificial magnetic field. The birds reversed the direction they perched whenever the magnetic field was reversed.

Several species of nocturnal migrants such as the Savannah Sparrow, the Blackcap, and the Garden Warbler have been shown to have a built-in magnetic compass, akin to the primitive compass used to calibrate the necessary aspects of sun and star compasses. Some birds may have the ability to detect a magnetic field because of minute magnetite crystals found in the animal's head. It is not known precisely how these crystals work, but similar ones have been found in living creatures that run the gamut all the way from lowly bacteria to *Homo sapiens*.

Research on bird navigation continues to be one of the most exciting, rapidly shifting aspects of ornithology, with many new and perhaps startling discoveries on the horizon.

THE ULTIMATE AIRPLANE

And so birds fly, often using the paths traveled by thousands of generations of their ancestors. Seemingly single-minded in purpose, they crisscross the continents.

Some fly at breakneck speed for short distances, like the Ruddy Turnstone, which travels more than six hundred miles a day and may reach its destination in a mere four days. Then there are those like the Arctic Tern, a medium-sized bird that takes its time, flying less than one hundred miles a day, feeding as it goes, on a trip that takes 114 days to accomplish. And there is the Albatross, a very large, long-winged bird that soars over oceans and spends as much as 90 percent or even more of its life in the air.

Large and small, fast or slow, whether they fly for months or days, these birds are aerodynamic wonders, their prowess unmatched in the kingdom of all living things. It is difficult to imagine that anyone who watches them with even the slightest curiosity could fail to come away without a sense of awe and respect.

No doubt the team of researchers bent on investigating whether the Gray-cheeked Thrush migrating through central Illinois used the stars to navigate came away ever so slightly humbled.

The team captured a thrush one afternoon and attached a tiny radio transmitter to it. The bird flew off at dusk on the next stage of its northern journey, followed by the ornithologists in a small plane. During the night, a severe thunderstorm and a rapidly emptying fuel tank forced the plane down. Yet onward flew the thrush.

After the plane was refueled and the weather had improved, the ornithologists continued their journey. The scientific team regained contact with the bird in Wisconsin, where it had landed at dawn, having flown right through the storm, without stopping once.

A hovering hummingbird's wing actions. The interval between sequential positions is four milliseconds. At this interval, this drawing would have to be repeated more than seventeen times to cover the bird's flight movements for just one second!

HOW DO BIRDS FLY?

The airplane has yet to be invented that can equal the flight efficiency of the most ordinary bird.

Consider the American Golden Plover. This adept flyer fattens itself up in the autumn and then journeys off across the Atlantic on a nonstop flight of 2,400 miles to South America. On arrival, the bird weighs about two ounces less than it did when it started the trip. Compare this to a small airplane, which normally would require a gallon of gas for every twenty miles. To equal the fuel efficiency of this bird, that same airplane would have to stretch a gallon of fuel to last for 160 miles!

How have birds managed to do what man can only dream about?

The two most important components of flight are high power and low weight. And evolution has been kind to birds on both counts.

First, birds are extremely light. For example, the skeleton of a pigeon accounts for only 4.4 percent of its total body weight, while a rat's skeleton is 5.6 percent of its body weight. Many of the bones in a bird's body are hollow. And the wings and tail – extremities critical for flight – are mainly formed by feathers, one of nature's lightest yet strongest materials.

Anyone who has watched an airplane taxi out onto a runway, suddenly pick up speed, and within seconds lift itself into the air, has witnessed firsthand the importance of power. A bird, too, is built with the power needed to lift it off the ground. Important components of this power are the bird's powerful breast muscles, which may account for almost half of body weight in the best fliers. Moreover, birds have a high metabolism (compare the temperature of some thrushes, 107 degrees Fahrenheit, with our own 98.6 degrees), so their engine is always "revved up" and ready to take off, with no runway necessary. A bird's heart weighs six times more proportionately than a human's, which allows birds a much greater cardiac output than that of mammals – essential during the strenuous exercise of flight. The bird's lungs, connected with air sacs that extend over much of the body and even into spaces in the bones, are structured so that oxygen can be absorbed continuously, whether the bird inhales or exhales.

A bird is designed so that body parts unrelated to flight are either minimal in size, or compactly located near the center of gravity. One example is the bird's reproductive system. Most of the year the sex organs of both the male and female atrophy so as not to burden the bird with excess bulk that would make flying difficult.

To further aid in its flight, a bird is also born with its center of gravity between and below the wings. To help center the weight, the head has been reduced by eliminating teeth and jaws. To aid in the breakdown of food, the bird evolved a gizzard in place of teeth, and the large gizzard is located near the bird's center of gravity. Efficiency of flight is the hallmark of birds, and adaptations such as these occurred early in their evolution.

But perhaps the bird's greatest and most distinctive assets are its feathers. A bird's feathers serve to keep it airborne, to insulate it against cold temperatures, to make it more streamlined and, incidentally, to keep its weight down. Because of feathers, a bird can not only live but thrive in parts of the world too cold for any other animal – provided food is available to keep its lifeblood flowing.

8

Building a Nest

In the midst of a sandy mound deep in the Australian outback, the Orange-footed Scrub-fowl lays a single egg in a large pit her mate has dug. The mother bird then covers the egg with sand and leaves. That task complete, she walks away, never to return to witness the birth of the chick she and her mate have created. Meanwhile the male tends the mound, regulating its temperature by adding more material to retain heat or kicking it off to cool the precious egg.

On the other side of the globe lives the Northern Oriole, a species that has been called America's greatest nest builder. With strong plant fibers for the nest cup and grass and hair for its lining, the female creates a deep, purselike structure hanging from the tip of a branch. Using her bill like a needle and the materials like thread, the bird meticulously weaves the nest, a task that observers have estimated may take four or five days. When she has finished, the oriole's nest hangs by about two hundred strands of fiber, capable of bearing a load of eight pounds or more, a strong and elegant home for the brood it will soon contain.

These two examples illustrate the incredible diversity among birds and the nests they build or, in some cases, don't build.

Birds build nests primarily for protection. A bird protects the nest itself by concealing it so that predators cannot find it or by building it in a spot inaccessible to enemies. Not only does the nest afford the eggs and the chicks that ultimately hatch from those eggs protection from predators, but it also acts as a barrier against inhospitable weather and assists in incubation by holding the eggs in place so that they don't roll about.

A warm dry nest, tended by nurturing adult birds, accelerates the maturation process, allowing the young birds to leave the nest sooner than they otherwise would. This is critical in ensuring the survival of the species because it is during this phase of a bird's life that it is most vulnerable.

Although we associate nests solely with breeding, some species of grass finches construct roosting nests, similar to their domed breeding nests but lacking the grass-woven tunnel found in the nests of breeding grass finches.

TYPES OF NESTS

The type of nest a particular species builds is determined by heredity, the raw materials available at the time the nest is under construction, the nest site, the experience and ability of the builder, and possibly by imitation of older birds.

As a general rule, the small perching birds known as passerines, which include more than half of all living species, build the most elaborate nests. This stands to reason because passerine species lay eggs that are small and fragile, and the hatching chicks are naked and unable to control their body heat.

Most modern-day birds incubate the eggs with their own body heat, as opposed to relying on the sun's warmth to hatch the chicks, a practice believed to have evolved along with the need to conceal and defend the eggs from potential predators and to allow more rapid development of the young.

The following describes some nests ranging from the simpler to the more complex.

Little or No Nest. It is believed that ancient birds early in their history used to lay their eggs on the ground like most reptiles. While most birds have evolved well beyond that stage, there are still some today who lay their eggs without building a nest.

Thus, the White Tern lays its single egg on the bare branch of a tree. The hatchling chick is equipped with very strong claws that enable it to cling to the tree.

The egg of the Ostrich, at three pounds the largest egg laid by any modern bird, is deposited in a scraped-out hollow on the ground, which may be home to several dozens of other eggs laid by several hens who share the same male. It is not uncommon to find an egg that has "wandered" away from this makeshift nest. Some speculate that the egg may be the first one laid (it may take two weeks for an Ostrich to lay all its eggs), or that two females came to the pit at the same time to lay and one kept the other away too long. The mass of eggs is incubated by the male at night and by one of the less conspicuously colored females during the day.

One of the most interesting birds that does not incubate its eggs in a nest is the Emperor Penguin, a bird that has managed to survive despite the rigors of its native Antarctic climate. The female lays her single egg on the frozen ground. She then passes it to her mate and sets off across the ice on her way to the open sea to feed. The male places the egg on his feet and covers it with a flap of skin that hangs from his belly. This is how the egg is incubated – in temperatures that can drop below minus 32 degrees Fahrenheit – for sixty days, during which time the male bird lives off his ample fat reserves. The well-nourished female then returns to relieve her mate prior to the hatching of the chick.

Ground Nests. Birds nesting on the ground are particularly vulnerable to predators and adverse weather, especially flooding. In response to the first hazard, many ground nesters and their eggs are colored so that they blend in with their surroundings, concealing them from enemies. In many cases, ground nesters tend to live in open or uniform habitats where the threat of mammal predators may be lessened. If the nest is flooded, it is deserted and the parent birds lay another clutch of eggs.

The majority of waterfowl build their nests on the ground. The duck builds a simple nest of grass in a scrape it makes in the thick grass, usually under a bush above the water's edge. For warmth, the nest is lined with down and the bird's own feathers.

Different indeed is the Wandering Albatross, a bird that takes great pains in the building of its nest. In treeless South Georgia, a cold island southeast of the Falklands, near Antarctica, the huge albatross pair first scrape a circular trench with their beaks, filling the center with soil, moss, and grass, heaping more and more material on the nest, and then trample it, until it is a solid, peatlike mass. Shaped like a small volcano, the nest measures three feet across and is three feet high, with a shallow cup at the top for the birds' single egg. The height of the nest protects the young bird from floods, and, thanks to the sitting parent's warmth, the egg or youngster is snug, despite frequent rains, fog, and even snow.

Adélie Penguins also build ground nests, using small stones placed directly on the frozen Antarctic ground. The stones apparently lessen the chance that the eggs will be buried by snow during a blizzard or flooded during a thaw. One bird, in fact, was observed during a thaw in which a steady stream of frigid water was running through the nest. With the eggs half covered with water, the male penguin was seen busily collecting and arranging the stones around him. By the next day, the eggs were above water, although the nest itself was surrounded by the stream. The eggs eventually hatched.

Mud Nests. Birds that use mud in their nests either scoop it into a mound or carry a mouthful at a time to the nest. Species such as the Australian Magpie-lark and the Willie Wagtail build bowl-shaped nests made from mud reinforced with bits of sticks, grasses, feathers, horsehair, and fur. While these mud-building birds construct their quarters on tree limbs, there are some mud-nesters who prefer the comforts of home – yours, perhaps. Swallows and American martins are quite happy to take over a barn, garage, or shed. Swallows generally attach their nests to a beam in the roof or against a wall, while martins seem to be partial to building their nests beneath the eaves.

One intriguing nest is made by tropical American oven-birds, some of which construct intricate mud nests, each with an entry hall that twists and curves its way to the nest. When dry, these nests with adobe-like walls are as hard as rocks and extremely difficult to break into.

Nesting in Holes. Many bird species make their home in a hole, in a tree, in the ground, or under a rock. Some birds such as woodpeckers actually create their own hole, while others seek out an existing opening. Some build little or no nest in the cavity; others construct elaborate chambers.

A hole nest is generally considered among the most secure because the nests are well hidden and easily defended against small predators. When a predator does stumble onto a hole nest, it usually is unable to kill or get past the adult bird.

The two-chambered mud nest of the South American Rufous Ovenbird. The bird builds the nest with a mortar made of sand, wet mud, and cow dung. Ovenbirds breed only when it rains, and cows are not always nearby. The drawing at right shows a vertically sectioned nest with the entrance at the right, partially separated from the egg chamber at the left.

This two-nest complex, connected by a tunnel, was built by another South American oven-bird, the Rufous-breasted Castle Builder.

Aside from the safety factor, hole nests usually maintain a moderate temperature and are cooler in the day and warmer at night than open nests. Rain and wind also are less of a threat to the hole-nester.

Hole-nesters are a varied group. There are the shelducks who may nest in rabbit burrows. The Crab Plover burrows deep into the dunes of the Red Sea shores. The beautifully plumed Kingfisher builds its tunnel in a river bank by repeatedly attacking the bank with its bill until an indentation is made. The bird then clings and perches, and begins to dig, shoveling the loose soil backwards with its feet until a tunnel is formed. The tunnel is just wide enough for the bird, which has to back out from its digging sessions until it makes a widened chamber at the end and finally has room to turn around. The eggs are laid in the chamber on a pile of fish bones and pellets, and while the chamber itself is relatively clean, the tunnel to it is often filled with rotting fish and foul-smelling debris.

Woodpeckers are probably the best known hole-nesting birds. Most woodpeckers, although not all, make their homes in a tree. Once a year the woodpecker pecks out a new home, hammering away up to six hours a day with a bill that is as effective as any chisel. To aid in its task, the woodpecker is equipped with strong feet to grip the tree and stiff tail feathers to help brace its body.

While the tree or branch under a woodpecker's assault may appear to be alive, most are, in fact, dead. An exception is the live pine trees for which Red-cockaded Woodpeckers have a particular penchant, although these more-or-less healthy-looking trees almost always have a fungus-rotted center.

Open Tree Nests. When most people think of a bird's nest, they envision a small cup-shaped creation of twigs, grass, and other fibers resting on the branch of a tree. While the size and construction of such nests may vary, this cup configuration is found in the nests of birds ranging from the tiniest hummingbird to the majestic eagle.

In shaping the cup or hollow, many species use the following method. First, the bird presses its body into the nest mold for about four seconds, then rises for five seconds, and repeats the motion again. With each pressing, the bird rotates its body from the previous position. The process is repeated ten or so times, during which the bird has made two or three complete rotations. The next time it brings more material to the nest, it may perform the turns in the opposite direction.

Few birds are as painstaking about their nest-building as the hummingbird. One species, the Ruby-throated Hummingbird, after choosing its nest site, plasters the area with saliva and then begins to stick the nest to its support. The nest is constructed so that its tight walls guard against heat loss. Not only is the nest difficult to spot because of its size, but it is camouflaged with lichens and blends in with its surroundings.

Another open tree nester adept at camouflage is the Australian Varied Sittella. The bird constructs its nest on the fork of a branch using cobwebs, bark, and vegetable matter. After the cup is built, the bird then decorates the outside with chips of bark, making it difficult to differentiate the nest from the tree. The incubating bird sits as low as it can to break its profile and prevent detection by a predator.

The American Bald Eagle builds its large platform nest of branches and twigs in the top of a tall tree. Called an eyrie, the nest may take several months to build and is used for many years, with the bird adding new sticks as renovation is needed. These nests can become huge, with the largest one on record measuring twelve feet in height and weighing over two tons.

Enclosed Nests. Some birds – especially those that live in tropical climates – build nests with dome-shaped roofs. While a roof is no guarantee against all predators, it does offer some protection against the cold and rain, helps prevent heat loss when the parent birds are away, shades the young birds from the scorching sun, and helps to avoid detection by predators.

At its simplest, an enclosed nest is merely a cup-shaped dwelling with a roof; at their most elaborate, such nests may have tunnels, ledges, and hidden entrances.

Probably the most intricate enclosed nest belongs to an African bird called the Hamerkop. The breeding pair begin with a pile of sticks in the fork of a tree and end up with a huge inverted pyramid. Sticks, grass, mud, and animal dung make the roof waterproof. So strong is this roof that it has proved capable of supporting a man's weight, and is often used as a nest platform by certain owls and Egyptian Geese, which move in when the Hamerkop family moves out.

Six or seven weeks after the first stick is placed, the nest is finished. An estimated eight thousand sticks and bunches of grass have gone into its making.

Hanging Nests. Suspended from the branches of trees on cables made of grass or spiders' silk, these creations are among the most wondrous of all avian architecture.

Like the enclosed dwellings, hanging nests are found more often in the tropics where the likelihood of encountering tree-dwelling predators is greater.

Although all these nests are the work of master architects, the Hermit Hummingbird of Venezuela also has considerable engineering acumen, as is demonstrated by its nest. This nest hangs from threads of spiders' silk. But since these are attached to a single point on the rim of the nest cup, the nest would seem likely to tilt dangerously. It doesn't, though, because the bird uses little lumps of clay or pebbles fastened below the point of attachment, which act as a counterweight, keeping the nest level.

Perhaps the most famous bird builders are the weavers, of which there are about one hundred species. Most weavers' nests are small globe-shaped structures, with entrances either at the side or below the nest, and usually with an attached tunnel. Some nests have hidden chambers and false entrances, in an apparent attempt to outwit predators. Using mainly their bills, weavers are able to knot, twine, and weave strips of grass to form the walls, floor, and ceiling of the nests, a process that can be completed in as little as a day. The ability to speedily construct a nest is crucial because a male may build several nests before it pleases a female.

Colonial Nests. A small percentage of the world's birds build their nests in close proximity to other birds, which are usually, but not always, of the same species. These colonies may consist of a few dozen nests or, in some extremes, may hold several thousand birds.

The majority of these colonies are made up of individual nests placed close to one another. A few species, however, build compound nests, a kind of apartment building for birds. The largest nest of this type belongs to the social weavers, small African birds that build a formidable complex. Every bird in the community works together to construct the large thatched dome of stems and grasses. Then, on the underside of the roof, each pair builds its own nest with its own private entrance. In

Various types of African weavers' nests (clock-wise from top left): nest built by a Gray's or Blue-billed Malimbe; weaver's nest with an entrance tube (many weavers build such a nest); and a kidney-shaped nest.

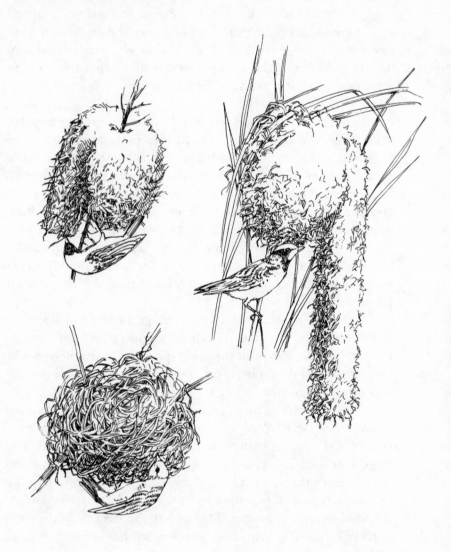

some colonies, several hundred nests are constructed under a single roof.

CHOOSING A NEST SITE

Birds – adaptive creatures that they are – have been known to nest in places that can only be described as bizarre. Among the locations: a wagon on a cross-country jaunt; the cistern of an outhouse; a much-used rowboat; a street lamp; a discarded kettle; and even a skull!

Despite these admirable feats of improvisation, most species, given a choice, opt for a more conventional nest site, whether it be in a tree, on a cliff, buried in the tall grass of a marsh, or in any of the other natural habitats that birds frequent.

The choice of a nest site is a vital factor in the continuation of the species, and in species where the competition for prime nesting ground is fierce, losers often pay a steep price.

The Razor-billed Auk of the North Atlantic seas and shores is one such bird. For this species, prime real estate equates with being on the ledge of a cliff (the ledges must be nearly horizontal) and the competition for the limited number of such sites is keen. The losers end up building their nests in more vulnerable spots, usually on or near the cliff. In the end, it is the cliff-ledge dwellers that are able to rear more young to maturity because their location makes it easier to escape the gulls who attempt to snatch the chicks' food out of the parent birds' mouths, as well as snatching the chicks themselves.

Sometimes a bird is forced to nest in a totally unsuitable spot because the prime sites are taken. The Magnificent Frigatebird prefers nesting on top of a cactus thicket. But a dearth of cactus in some island areas has forced the species to nest instead in the tops of bushes and trees. Occasionally, the birds lose their balance, fall, get caught by the neck in the fork of a limb, and hang there until they die.

Many birds have evolved so that their nest site is purposely placed near the nest or burrow of a larger or hostile animal which helps indirectly to ward off predators. In some cases, the nest may be built near or even within that of a larger bird that may or may not share the same enemies. Various waterfowl that are preyed upon by the Arctic fox tend to nest close to Snowy Owls – the fox's most formidable enemy. In doing so, less aggressive birds enhance their chance of survival. Their young may be at risk after hatching, but they then quickly move to water, well away from the owls.

As a way of protecting themselves, some species over the generations have taken to building their homes near insects that most creatures would go out of their way to avoid. South American caciques construct flask-shaped nests near the nests of wasps. And in eastern Africa acacia trees are the shared homes of Gray-headed Social Weavers and ants.

Some birds carry the relationship one step further by excavating a nest hole within an insect nest. Certain species that inhabit the tropics are known to nest within termite mounds or even in carton-ant nests in trees – the nesting cavity promptly is sealed off by the ants or termites and the birds carry on with their nesting.

In these examples, the insects don't seem to bother the birds, although they are quick to attack if another creature enters the tree. In some cases, an added benefit to the birds is that the insects are a convenient food source, although usually birds prefer not to feed close to the nest, so as to avoid attracting the attention of a potential predator.

While it would seem that a bird's power of flight provides it with limitless nesting possibilities, several factors intervene to narrow the scope of those choices. The bird's size and the structure of its bill and feet are important. A bird such as a swallow with its weak, wide bill would have as difficult a time chiseling a hole in a tree as a woodsman would have cutting kindling with a table knife.

Then, too, there are ecological limitations that restrict where many birds can best rear their young. With its need for strong winds to help lift it off the ground, the Royal Albatross builds its nest on a clifftop so that it is within a few steps of the strong updraft it needs to take flight. To build anywhere else would be tantamount to crippling this bird.

When it comes to selecting the actual spot on which to build the nest, in most species both male and female participate to varying degrees, although the female usually has the final say. There are many exceptions to this, of course. Male ducks and geese, hummingbirds, and birds-of-paradise generally neither help find the nest site nor build the nest, whereas the males of some migratory species such as the House Wren arrive at the breeding ground before the female and begin building a nest before the mate arrives.

Finding the all-important nest site is not always accomplished quickly. A female European Chaffinch scouts among the trees and bushes, flying from one to the other. Finally, she will land on a branch and hop down it until she comes to the fork. If there is a deep fork of two or more branches, she will turn around and around, flicking her tail and looking at it. She will then hop into the fork and ruffle her feathers. She will inspect several sites in this manner, even bringing building materials to some, until she finally decides on the best place. Presumably, there is something important about the branch configuration and strength that even an inexperienced European Chaffinch is able to recognize.

The female Australian grass-finch does not herself seek out potential nest sites but relies on her mate to do so. Once the male finds a site, he attracts the female by hopping back and forth and calling out. The female then flies to inspect his choice. If she doesn't find it suitable, she simply flies off and the male continues his search, a quest that may not be successful until one hundred or more sites have been rejected.

Some birds like the European Mistle Thrush show anticipatory behavior when choosing a nest site. The female may sit on or "brood" a bare site for two or three weeks before she actually begins construction of the nest.

NEST-BUILDING AND COURTSHIP

Males of some species use elements of nest-building in their courtship of females. The degree to which this is done varies greatly. In some species such as the grass-finch just mentioned this may be as rudimentary as the male bird picking up and holding a piece of grass or feather in its beak in front of the female. Other males build one or more (up to ten in some wrens) nests, complete save for the lining, which is constructed by the female when and if she accepts the nest. In some weaver species, these extra nests come in handy, allowing the male to mate with two or three different females if it is able to defend all the nests against nearby males.

The male Green-backed Heron upon arriving home in the spring takes over an old twig nest, which it repairs or uses as the basis for a new nest. The bird calls attention to itself by loud calls and by making a snapping sound with its head and jaws. When a female approaches, the male is aggressive initially, but she keeps circling back. This eventually reduces the male's aggression and he becomes more sexually aroused and allows her to enter the nest. She then begins moving twigs, which causes the male to pass more twigs to her. The male then begins gathering still more twigs, which he brings to the female. They then complete the nest together.

If flowers can work for the human, why not the bird? They do for the male Black-throated Weavers of India. This bird decorates his nest with flower petals, using wet mud to paste the petals to the nest. He then attempts to attract available females to his colorful place.

Male bowerbirds have nothing to do with a nest. The females mate and go off on their own to raise the young. But the males construct often elegant "bowers" of sticks and other vegetation and decorate them with colored stones, flowers, and even the juices of some colorful berries in order to attract females for breeding.

One bird whose courtship and nest-building practices are closely bound together is the Village Weaver, a common colonial species in Africa. The male birds usually arrive in the area before the females, vie for a small territory within the same tree, and then set to work weaving intricate nests of grasses and palm fronds. When an unattached female is near, each male attempts to attract her by hanging upside down from his nest and flapping his wings. Once a female enters his nest, the male begins singing to her. Then, like an Army sergeant inspecting a recruit's bunk for perfectly cornered sheets, she begins pulling and tugging at the nest, as though testing the tightness of its construction. The female may visit several nests before deciding on one. Once she makes her choice, she arrives with a bill full of grass or other material, sets to work lining the nest, and within a day or so copulates with the male owner.

What impels her to make her particular choice? It appears that important factors are the number of males nesting in the tree (studies have shown that colonies with fewer than ten males are less successful than larger ones); the frequency of the male's wooing with his upside-down flapping display; the freshness and strength of the nest itself (as the nest ages, it browns and becomes weaker and thus less appealing to the female); and the nest's inaccessibility to predators.

Sometimes a male's nest fails to attract any female. If the nest fades from green to brown without having enticed a mate, the male bird tears it down and begins anew. What makes a bird destroy something it has worked so hard to build? One factor may be the small number of good nest sites. If one nest isn't successful in attracting a female, the male may not have the option of looking elsewhere. Color may also be important. In experiments in which the color of the nests was manipulated – fresh nests were painted brown, while nests the same age were sprayed green – researchers found that the birds were three times more likely to destroy the brown nests.

NEST-BUILDING BEHAVIOR

Snatched from their outdoor aviaries on the campus of the University of California, Los Angeles, the group of week-old Village Weavers were housed in cloth-lined bowls in an attempt to discover the basis of their extraordinary nest-building ability. The birds, taken from their nests before their eyes were even open, were denied access to nest materials. When they outgrew their bowls, they were moved to cages.

Two other groups were used in the experiment. The second group were only partially deprived of building materials and were reared in Village Weaver nests until they started to fly, at which time they were transferred to cages. The third group – the control – had normal access to building materials and were left in the aviaries under the tutelage of their parents.

Village Weavers in the weaving of their intricate nests use fresh palm fronds and grasses. While the deprived birds lacked access to such vegetation, this didn't stop them from trying to weave. It was not unusual for one of the researchers to see a bird trying to weave its own tail feathers or wrestling with a neighbor in an attempt to weave its wing feathers. Moreover, these birds, like the control group, preferred the color green over yellow, red, blue, black, and white. Green, of course, is the color of the fresh grasses and fronds used to weave the nests. The more experienced the birds became, the greater their preference for green.

In order to weave a nest, the male Village Weaver must first be able to tear a ten- to fifteen-inch strip of grass. This is accomplished by perching on the stalk, biting through one edge of the grass, and then flying with it in the direction of its tip. This takes practice, and young birds frequently tear the blade in the wrong direction or tear the strips too short.

In the experiment, five male weavers who had been deprived of nest materials after the age of seven weeks were exposed to fresh reed grass when they were a year old. During the first week, the deprived group tried fifty-two times to extract the grass, while the control birds needed only five attempts.

Once the birds had learned to tear the strips, the partially deprived birds often tore strips too small for weaving. The birds improved quickly, but after three weeks of practice, they were still not equal to the control group.

In its natural habitat, a yearling Village Weaver builds a nest that is crude, compared with what he will eventually be capable of creating. To determine the significance of practice, three of the deprived males and three controls were tested for their ability to weave. During the first week of testing, the deprived birds didn't weave, while the controls wove well. The second week the first group wove a few stitches. For the first few weeks, the deprived group wove significantly less than the birds who had been reared around building materials. After three months, two of the deprived birds had managed to weave two nests, while the three controls had built eleven nests, the quality of all the nests being equal.

When retested a year later, the two groups were equally adept at weaving, with the exception of one bird from the deprived group. This bird rarely got a chance to weave because every time he tore a strip of grass, one of the other birds stole it. The result of this two-year deprivation was that the bird never learned to make a nest. While he continued to tear strips of grass and attempted to weave, he couldn't seem to get it right.

Most features of nest-building appear to have an instinctive basis. The young megapodes who hatch from their underground nests and scurry off into the brush, never to meet their parents, nevertheless are able to build nests typical of their kind when they come to breeding age. No adult bird teaches or instructs by example. The young birds simply know what to do.

However, as the Village Weaver experiments and others show, learning also plays a role for many, especially social, bird species. Practice and the bird's physical maturation combine to hone the basic instincts.

Not that those instincts are always infallible. When prevented by its mate or some other reason from brooding its eggs or young, a gull commonly begins fetching building materials. Rarely, the bird will accumulate so much nesting material that the eggs or newly hatched chicks are suffocated. Wilson's Petrels, a species found in Antarctica, have been observed in

*Various knots and stitches used by weavers in
nest construction.*

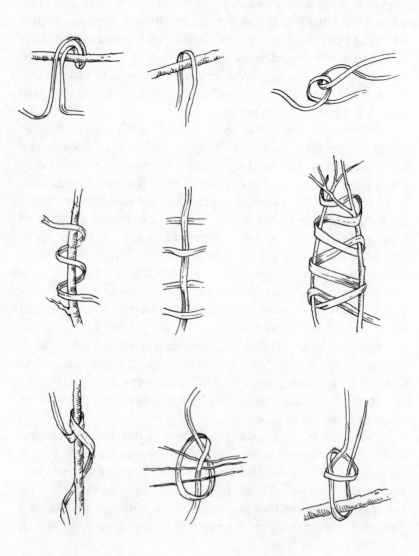

similar nonproductive behavior. When moss was put in front of these birds' nests, the petrels stuffed their nests with so much that there was barely room for the birds. Often such behavior is a function of youth and lack of experience.

Regardless of age or experience level, however, it seems that when a bird's nest-building instinct is triggered, the desire to build is so strong that almost nothing will stop it. To determine what, if anything, would halt a Red-winged Blackbird's nest building, eggs or young chicks were put into the unfinished nest. The bird ignored the contents of the nest and continued to build.

As experiments with the Prairie Warbler suggest, constructing a nest is a central experience in a bird's life. In the warbler study, uncompleted nests were replaced with finished ones. Yet the females pressed on, continuing to build. Even when nestlings were placed in the nests, three of the females simply ignored the chicks and continued with their task.

One bird, though, stopped building and tried to eject the interloper from her nest. When that failed, she did what adult birds have been doing since they first evolved. She began to feed it.

BIRD'S NEST SOUP, ANYONE?

For those who have wondered if there really is a soup made from a bird's nest, the answer is, yes, indeed there is. Not only does bird's nest soup exist, but it is, to some peoples in the Orient, quite a delicacy and is served in gourmet Oriental restaurants throughout the world.

Take one swiftlet. If you're inclined to give it a try, you should know that the only birds that produce edible nests are three varieties of swiftlet, all found in southeast Asia. Virtually all commercially harvested swiftlet nests are from a single species, the Edible-nest Swiftlet.

Swiftlets reside in deep, dark caves. The largest colony of the Edible-nest Swiftlets is in the Niah Cave, a multichambered cavern in Sarawak, a part of Malaysia. One cavern alone in this labyrinth could easily hold St. Paul's Cathedral.

Did you say bird saliva? The swiflets build both "white" and "black" nests. Both types of nests are constructed from the bird's saliva, although the black nests also contain feathers and vegetable matter, both of which have to be removed before the nest can be used in soup. Understandably, the white nests are more costly.

During the breeding season, the swiftlet's salivary glands become enlarged, producing huge quantities of glutinous, almost white saliva. The nesting bird flies at the cave wall, lands, clings, and then deposits a drop of saliva with its tongue. The saliva quickly hardens. The process is repeated again and again until a horseshoe-shaped foundation appears. A shallow cup-shaped nest is then formed using the same method.

How brave are you, anyway? Swiftlet nests have been farmed for many years, although in recent times the number of nests have dwindled markedly. It's not that the swiftlets aren't holding up their end. The Niah Cave is estimated to hold 1.5

million nests in the breeding season, and when the nests are harvested at the correct time, the swiftlets rebuild.

No, the reason swiftlet nests are not as easily obtained as they once were is because humans, it seems, are less willing to climb into a dark cave with some 4 million bats for company. The intrepid soul who does take the risk must scale three- to four-hundred-foot walls in search of the prized nests. One slip and one's soup-making career, not to mention one's life, is over.

Was it worth it? Once you have your nest, place it in a bowl with some broth or other liquid, pieces of minced chicken, chicken liver, and/or vegetables. Or, you can opt to have it in a plain soup, in which case it is said to have no taste at all.

The Horned Coot, which breeds where there is little vegetation, builds up a mound of stones in the shallow part of a lake in the Andes Mountains, making an "island" nest and using aquatic plants to line it.

Two male European Yellow Wagtails in a territorial dispute. The birds puff out their bright yellow breasts and then begin swaying from side to side. The display climaxes with one chasing the other away or the two pecking and clawing at each other.

9

Courtship

The Kakapo, a huge flightless New Zealand parrot teetering on the precipice of extinction, doesn't court often. But when it does, it is an event that can be heard for miles around.

Kakapos, nonflying birds whose ranks have been reduced to around fifty, don't breed every year like the majority of birds. Rather, field work and studies suggest the male bird begins his elaborate courtship display only in years when he can predict that food will be plentiful for any chicks produced from the liaison. Thus, this unusual forecasting ability of this endangered bird means that in alternate years in which the environment cooperates to produce a bumper crop, there may be only a two-year gap between breeding seasons; in more austere times, the nocturnal Kakapo may breed only every fourth year.

When the time is judged as right, the male Kakapos gather together for a courtship competition known as a lek. Each male scratches out a bowl-shaped depression in the dirt, perfectly designed to fit the shape of the bird's body. The male, which often has increased his weight by as much as 60 percent in preparation for this event, then clears the area of twigs and weeds. Often, the "bowl" is positioned so that it is backed by a large stone, tree trunk, or shrub, props that help project the noise the bird is about to make.

Then it inflates its air sacs, swelling its breast and throat until its whole body is the shape of a large football. Slowly the Kakapo expels its breath, the result of which is a series of drawn-out, far-carrying grunts known as "booming." Although booming is seemingly soft, humans standing several miles away have reported hearing the noise. In one night of intense courtship, a single male has been known to make as many as seventeen thousand booms. The booming may continue all night, every night (weather permitting), for up to three months.

By booming, the male attempts to make himself more attractive than his neighbor, who is also booming to the best of his ability. Females, some coming from miles away, approach the lek. When a female comes within sight of one of the male's bowls, he begins to dance. From one foot to the other he rocks, making low clicking noises as he slowly advances toward the interested female. As the two come closer, the male begins to spread his wings, raising and lowering them like a butterfly in flight.

Turning as he dances, the bird parades his extraordinarily beautiful green and yellow plumage before the prospective mate. Then, with his wings spread wide, he backs toward her, wingtips touching the ground and head bent low. The bird stops and waits. The option is now the female's. If she approaches, the two will mate. If not, the male will repeat the display for another female.

Unlike many bird species in which the male helps the female build the nest and care for the young, the male Kakapo's only role in reproduction is that of fertilization. No long-term relationship is established between the male and female. After copulation, the female goes off by herself, lays her eggs in a hole, incubates them, and cares for the young that hatch. As for the male, he continues booming, hoping to attract another female with whom to mate. Unfortunately, in this species there are very few males and predators such as feral cats have greatly depleted and fragmented the populations, making the future of this unique and bizarre "boomer" rather doubtful.

COURTSHIP DISPLAYS

The male Song Sparrow takes no chances that the female of his choice will fail to notice him. After singing so persistently that she would have to be deaf not to hear his melody, the male abruptly stops. He then flies down at her, knocks her over, and flies away, singing at the top of his lungs.

This is but one of many courtship displays, each as individual as the species itself. Although reproductive behavior in general appears to be based upon building blocks of unmodifiable behaviors, there is great diversity in the way in which various bird species court and ultimately breed.

Birds have a natural reluctance to allow other birds into their personal spaces. When another bird does venture too close, a typical reaction is one of hostility. It isn't difficult to see how this would hinder a bird's chances of mating and producing more of its kind. Thus, nature has designed a way to bring the sexes together, overcoming their basic aggressive tendencies and hesitation to approach within touching distance, barriers that must be broken if eggs are to be produced.

In essence, a courtship display is a necessary prelude to mating and the raising of young. Through a more or less elaborate set of courtship rituals engaged in by both sexes, an appropriately stimulated male and female will pair, with the female generally making the final decision. This makes perfect sense when one considers that a male bird produces tens of millions of sperm, millions more than will ever be needed. Yet a female's reproductive life is more limited, making each egg an important investment. It is little wonder that a female shopping for a mate uses various cues such as looks, ability to provide – as evidenced by quality of territory – and, in some species, a bird that knows how to build a nest that can withstand a monsoon. She may even choose a male based upon where he is positioned in the lek. The cues the female uses are those selected in nature to ensure that she mates with a male with good genes.

An unmated male Common Black-headed Gull has a pairing territory where it waits for other gulls to fly past. Apparently, the bird does not differentiate between male and female passers by because there is no difference in color between the sexes, so it makes aggressive calls and displays, which are ignored by the passing males. A female in search of a mate, however, may land.

At first the birds are fearful and aggressive toward each other as they begin courtship displays; sometimes the male is overly aggressive, causing the female to fly away. Often, though, the male will direct his hostility toward other male birds or even peck at the ground, ripping grass as though it were an enemy's feathers. Gradually, the male's posture changes. He no longer uses the aggressive upright stance so frequently and changes to a hunched position, with the head turned away. This conceals his brown hood and dangerous beak, encouraging the female to approach more closely.

Before a female gull chooses her mate, she may visit many males, spending time in a male's territory, leaving, returning, traveling back and forth several times before she makes her choice. Each time she enters a male's territory, the two are initially hostile, although the hostility is not as strong as in the beginning, and the male no longer attacks. Slowly, they get used to each other, gradually leaving the pairing territory for a nesting territory, where they will mate and build a nest.

Most birds breed annually, so for most of the year the bird's reproductive organs are relatively undeveloped – essential when you consider that in many species the male's gonads expand to two hundred times their normal size during mating season. (In many tropical species, the gonads remain somewhat developed even after breeding.) Thus, courtship is important in synchronizing the readiness of male and female to mate and in stimulating ovulation. Not only is the female stimulated, but the displaying male bird also becomes sexually excited.

Courtship in the avian world encompasses many different elements. For many birds song, important in territorial establishment and defense, plays a major role in attracting and wooing a mate. Some birds sing more at the start of breeding, continuing even throughout the nesting period. Others, like the Sedge Warbler, are silent once they have found a mate. Songs and calls are not the only vocal tools used by birds during courtship. The tom Turkey rattles his quills, while woodpeckers will drum on dead limbs and occasionally on tin roofs to attract and stimulate a potential mate.

Courtship feeding occurs in many species, usually after the two have paired and mated. A male Malachite Kingfisher will present his female with a fish, while a Eurasian Bee-eater will offer his a dragonfly. While typically the males do the feeding, some hummingbirds are unusual in that the female bird brings food to her mate.

Scientists have shown that the inclination to feed during courtship is an instinct with which a bird hatches. In one study, Red-crested Pochards were taken from their parents after hatching and hand-raised. When they matured, the young ducks, exactly like their ancestors, performed the courtship-feeding ritual of diving for pieces of plant, which they then presented to females prior to copulation.

Although the extra food the male bird bestows upon his mate no doubt helps her meet her nutritional needs during egg-laying and incubation, the main function appears to be that of cementing the bond between the pair. The female European Robin may beg her mate for food even if she is surrounded by worms.

Males of some species will feed their mates just prior to mating, indicating that courtship feeding may trigger coition. A male Snail Kite presents a female with a snail. The female, in turn, removes the snail from the shell and eats it. No sooner has she swallowed the snail than the male mounts her. The male Roadrunner, on the other hand, holds a mouse or lizard in his beak during copulation, feeding it to the female only after they have mated, as if he were offering her some sort of reward for her participation.

In some species of pigeons, there is no actual feeding but rather a ritualized "billing," in which the pair touch bills or the female puts her bill into the male's empty mouth.

Pairs of various crows, herons, pigeons, parrots, terns, and albatrosses frequently preen the plumage of a prospective mate as part of the courtship ritual. From a practical standpoint, all preening is useful in that the bird is cleansed of the ticks, mites, lice, and fleas that cling to the feathers of the head and neck, areas the bird can't reach for itself. Again, however, it appears

To determine whether female Long-tailed Whydahs of Africa prefer males with long tails, the
tails of some males were shortened, while the
tails of other male whydahs were lengthened.
Control I males had their tails cut off and
then restored, while Control II birds had unal-
tered tails. The experiment showed that the
ability of a male Long-tailed Whydah to
attract a female was directly related to the
length of its tail (the longer, the better).

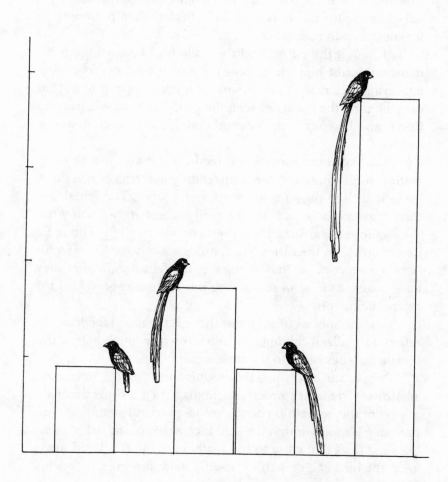

that this reciprocal grooming serves to allow contact and lessen aggression, as well as to synchronize the paired birds reproductively.

Some male birds are dazzling to look at, and these birds use at least some aspects of their colors to their advantage in courtship. In fact, there often is a direct correlation between a species' courtship movements and its showiest or (to us) most beautiful attributes.

The male Green Peacock, a stunningly beautiful bird, spreads his fanlike tail while approaching the female. Then at just the right distance, the bird dramatically shakes and quivers, every quill rattling, dazzling the peahen with his vibrating jewel-like plumes. The male then turns around to show his drab rear, momentarily cutting off her view of his plumage, before he again faces her and continues his display. The male then screams what might be interpreted as a bird's rendition of the Tarzan jungle call to attract her.

Few birds can rival the beauty of the birds-of-paradise, some forty-two different species, whose feathers represent more colors than can be found in a child's deluxe set of crayons. Not only are the males brilliantly hued, but many sport elaborate head decorations, which they use to attract potential mates. The Six-plumed Bird-of-Paradise is one such elaborately-coiffed bird. Projecting from the male's head are six barbless feathers, each tipped with a small oval vane, giving each one the appearance of a racquet. During courtship, the excited male may swing these racquets forward, dangling them toward the object of his desire.

"Dance" rituals are not uncommon among courting males. The legs of the Black-crowned Night-heron develop a bright rosy tinge during breeding season. If a female approaches, the male lowers his head with one cheek parallel to the ground, while uttering a guttural call of greeting. The bird then raises his head, at the same time contracting the pupil of his eye so that the eyeball actually protrudes from its socket, exposing the

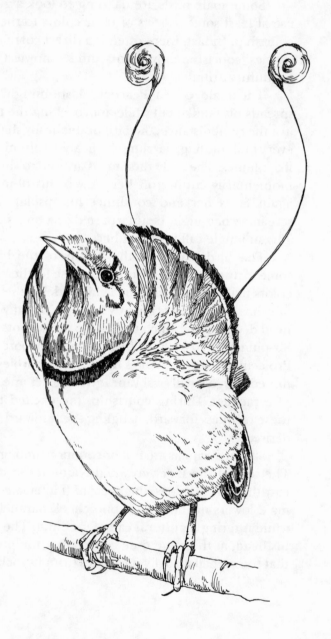

The male King Bird-of-Paradise displays his beautiful purple, red, white, green, orange, grey, and brown plumage to attract females.

bright red of the iris. The bird's plumes stand straight up and may even fall over his face as he bows to the female.

In bird societies where courtship is communal, dances may be performed by pairs of males or even entire groups. In a tropical rainforest, a group of male Yellow-thighed Manakins sit, each on a separate limb, awaiting the approach of a female. When a female does come close, the group swings into action. One male first stretches high on his legs to show off his bright yellow thighs, holding his black body horizontal and his scarlet head downward. With the wings partially open, each bird then reverses his position on the limb in rapid sequence, first facing forward and then backward, giving the appearance of a wheel in motion.

Just because a bird doesn't live on land doesn't mean it can't dance. Many waterbirds perform their version of a dance both on and in the water.

The male and female Great Crested Grebe swim toward each other on the water's surface until their beaks touch. They then dive under water, emerging face to face, each carrying a small piece of water weed in its beak. Rising from the water, the grebes' pristine white bellies pressed to one another, they weave back and forth. The male and female Western Grebe do a fantastic, synchronous rush across the surface of the water, side by side, at high speed and almost vertically.

Courtship also is conducted in the air. In some cases, a courting male bird does nothing more than fly in front of the female, making sure that his most brilliant colors or distinctive markings are positioned so as to give her the full impact of his presence.

In what has to be one of the most spectacular courtship flights, the male Common Lapwing rises slowly from the ground, speeding up as he rises at a steep angle. Suddenly, the bird plunges downward, turning and twisting, somersaulting, giving the impression he is about to crash. His wings produce a loud humming sound, as he pulls out of his dive and steers a straight course again.

In most species there are distinct differences between the sexes, a sort of badge that allows birds to quickly and easily identify potential mates as well as competitors or enemies. In some species this "badge" is barely visible. A human eyeing both a male and female Northern Flicker would find the birds identical save for a small black mustache-like mark on the side of the male's head. At the other extreme are species that have evolved so that the males and females don't look as though they were cut from the same mold. In sharp contrast are birds such as the Williamson's Sapsucker, a woodpecker species in which the sexes are so different in color and pattern that they were once thought to be separate species. Although there are exceptions, the males usually are the more colorful and brightly ornamented, the females more drab and colorless.

For some species of birds, a component of courtship is discovering who is who. In these birds both sexes usually play an equal role in courtship.

One group in which the sex is not readily apparent within the species are the penguins. To overcome the hurdle of sexual identification, the penguins have adopted a method of trial and error. A male penguin may place a pebble at the feet of another bird. If this second bird happens to be male, the encounter may erupt in conflict. If the second bird is female, however, she may not be interested and simply ignore the male. But if she is ready for the male's attention, she won't discourage her suitor. The pair will go through various rituals that include deep bowing to each other, stretching their beaks upward, trumpeting, neck-twining, and, ultimately, copulation.

MONOGAMOUS BIRDS

The Common Moorhen (a rail) is not so common when you consider the female's dominant role in courtship. In this species, it is the female rather than the male that initiates the process. Once she stakes out a claim to a potential mate, she is willing to fight for him, striking any competitor with her long clawed toes.

Not surprisingly, the larger female moorhens have their pick of the best males, with the less desirable mates going to the daintier birds. What is surprising is that the males in the greatest demand as mates are the most diminutive. Presumably they are so attractive because not only do they need less food than their heftier compatriots, but they can put on comparatively more fat, which means that when the time comes they can spend more of their day incubating the eggs and less time eating. Thus, a female with a small, plump mate can be working on her second clutch of eggs, while the first one is still being brooded by the male.

If you were somehow allowed access to the breeding grounds of the world's birds, you would observe a plethora of breeding habits. You would see some like the male weavers of the wooded grasslands of Africa who mate with as many females as possible, leaving the females to raise the offspring on their own. The female Northern Jacana, on the other hand, is the rare bird (fewer than 1 percent of birds) that practices polyandry. In this bird's reproductive world, the female sets up a territory, stays with one male for several days on one part of her territory, lays a clutch of eggs for him to incubate, and moves on to another male elsewhere on her territory. Her maternal duties completed with the laying of the eggs, the female doesn't visit the nests except to drive off predators or when one of the incubating male birds sounds an alarm.

Then there are birds such as the Tasmanian Native Hen (a large rail) that form breeding groups. In this particular species, two brothers and an unrelated female may stay together for life.

However diverse the avian world's breeding systems, monogamy is practiced by 90 percent of the birds on our planet. This doesn't mean necessarily that most birds "marry" and grow old together. Some mated pairs in some species do remain together for life. For others, monogamy may simply entail two birds coming together, raising a brood, and then going their separate ways, only to mate the next year with a new partner. Even ostensibly monogamous males will attempt to copulate with neighboring females if they come upon a female whose mate is not nearby. Many duck species are monogamous only until the female lays her eggs.

The prevalence of monogamy probably is due to the fact that rearing a brood of chicks is no easy task, particularly if they require feeding by the adults. Bird eggs tend to be large (relative to the weight of the parent). They need to be constantly incubated, and this often is best accomplished by sharing the burden. Once the eggs hatch, the work is often far from over, requiring two adult birds working virtually nonstop to care for a brood of helpless, hungry nestlings. As further testament to the idea that monogamy is born out of need, in the 10 percent of birds that reproduce using another mating system, only one parent is needed to successfully rear the offspring.

It is difficult to speculate on what holds a mated pair together. It appears that in species where there is a substantial delay between the time the birds pair and when they actually mate, the bond is longer lasting. Bearded Tits pair when they are just over two months old yet don't actually mate until they are one year old, while Laysan Albatrosses are together for two or three years before they breed. Mated pairs in these species probably stay together for life.

In some species who mate for life, there appears to be a strong tie between the pair. Twice, observers saw examples where the mate of a dying American Black Duck refused to fly out of gunshot range with the rest of the flock. Most albatrosses, petrels, Manx Shearwaters, geese, swans, Common Terns, Barn Owls, Adélie Penguins, and many parrots have long-term pair bonds, some of which last until death.

Territory is a factor in a pair's relationship in many species. The attachment to territory for the male White Stork is undoubtedly stronger than its devotion toward its mate. This male bird will fight for its nest (which takes a long time to build and is used every year) but not for its mate. Pairs of Wilson's Petrels, oceanic wanderers, return to the same burrow on the same island every year; presumably their fidelity to the site is more influential than to each other. These birds have a rather strong odor and may be guided to the correct burrow by that odor. As a result, the male and female both usually return to the same burrow even if they were apart prior to the breeding season.

A factor in pair faithfulness may be simply recognizing one's mate. In one experiment, several pairs of Ring Doves were separated after two breeding cycles. After a time – in some cases as long as seven months – each female was given a choice between a new bird and her former mate. Twelve out of fifteen females opted for the familiar male, most likely cued by its voice.

Staying with the same partner equates with breeding success in many species. This may be due, in part, to the experience of the older birds. Black-legged Kittiwakes winter at sea, and then return to their colony at the onset of the breeding season. The first to arrive are the older birds; a month or two later the less experienced birds arrive, and a month after that the breeding novices straggle in. As a result of timing, the experienced male kittiwakes and the experienced females pair up. The female kittiwakes that breed with their previous mate lay their eggs as much as a week earlier than those that team up with a new partner.

Among some species, failure to breed successfully equates with "divorce." In one study of kittiwakes in England, two-thirds of females who failed to raise a brood did not mate with the same male the next year, compared with one-third of the females who successfully bred after having mated with the same male the second year. In the case of Kittiwakes, breeding failures appeared to be related to the pairs' inability to synchronize shifts during incubation. As a result, the eggs were often left exposed and cold, which led to the death of the chicks.

Both partners in a mated pair have a major investment in the partnership. In many species, the female takes a considerable amount of time finding the right mate. Once she decides, the chosen male may be worse than any jealous human husband. Not one to share his prize, the male half of a pair goes out of his way to ensure that his female doesn't get close to another male, guarding his impending paternity as though it were some priceless treasure. As the female finch hops around, collecting material to build her nest, her mate is never far behind and ever watchful. As egg-laying time for the Black-billed Magpies approaches, the male guards his mate, keeping within five yards of her.

An observer once noted what happens when a male let down his guard. The male bird had the misfortune to doze off in the midst of guarding his female. A neighbor, spotting an opportunity, left his own mate and flew to the other female, attempting to mate with her. Unluckily for him, the groggy male woke up and was able to chase the interloper away.

Although males are zealous in protection of their own paternity, they aren't above trying to copulate with another bird's mate. To do so, in fact, may boost the number of a male's progeny. Actual "rape" is common in many duck species. While a male can usually drive off one offender, he is powerless against several drakes. As for the female who undergoes the attack, she may be half drowned in the process.

POLYGYNOUS BIRDS

For the female Lark Bunting, a grassland bird of western North America, being the primary female in a male's harem is the equivalent of the good life. It means essentially that the male takes an active role in the care and feeding of the female and her brood. So why would other female Lark Buntings be content to play second fiddle in the male's life, mating with him, living on his land, but getting little else in return?

Nestling Lark Buntings are particularly vulnerable to heat and they often die because of it. Thus, if a male has a good territory, meaning one with plenty of shaded nest cover, many females opt to raise their broods there by themselves, rather than pair with males that occupy less desirable territories.

The Orange-rumped Honeyguide has a similar situation. This species eats the wax and grubs from bees' nests. Naturally, the choice territories surround bees' nests, which are much less plentiful than honeyguides. Thus, a male lucky enough to control such an asset in his territory is a magnet for any passing female. One male was observed mating with up to eighteen different females in one season.

Birds who are polygamous mate with more than one partner of the opposite sex. There are two forms of polygamy: polygyny is the term used to describe a system in which the male mates with several females; polyandry occurs when a female mates with more than one male.

Only 2 percent of birds in our world are truly polygynous, although recent research indicates that males of many monogamous species do copulate with females other than their mate.

Polygyny most often occurs when the female is capable of caring for her offspring with minimal or even no help from the male. Generally, this means that the food supply is good and one parent alone is capable of gathering enough to feed the young. Precocial birds such as ducks, geese, and chickens are more apt to practice polygyny because at hatching they are covered with down and capable of being led to food, unlike virtually helpless altricial young, which require the efforts of two adult birds if they are to survive.

The importance of food in directing a bird toward one mating system or another is to be observed in England and Holland in the European Northern or Winter Wren. In these countries, 50 percent of all wren matings are polygynous, compared to 6 percent among American House Wrens. Yet on the isolated Scottish isle of St. Kilda, virtually all wrens are monogamous. Why? On this small island food is in short supply and it takes all the efforts of two adult birds if a brood is to reach adulthood. Conversely, the wrens that populate England and Holland are surrounded by abundant food stores, making it entirely practical for a lone female to raise the brood. In tropical areas where there is great year-to-year variation in the food supply, the pairing may also vary each year in some species.

In a typical polygynous system, a male will mate with several females, which incubate their eggs in separate nests and rear the chicks on their own, usually in the male's territory. This is not to say that all polygynous birds do not establish some sort of pair bond between mates.

In some species, polygyny results, in part, from the fact that there are more females than males. The Ostrich is one such bird. The male has a large territory, which he uses to attract a female. The hen scrapes out a spot for her eggs in the dirt and lays them. A few days later, hens that have not found a mate arrive, laying their eggs in the same depression. The first female lets the hens lay their eggs but only she and the male incubate them. It isn't unusual for the nest to eventually contain thirty or forty eggs, not all of which can be properly incubated, causing the incubating pair to push some to the edge. The advantage of this system, it appears, is that the primary hen is able to form a buffer between her eggs (which she recognizes) and those of other hens. Her centrally located eggs are more likely to hatch than those on the periphery and are less likely to be stolen by a predator.

POLYANDROUS BIRDS

The polyandrous birds are a study in sexual role reversal. This small minority of the world's birds – fewer than 1 percent – defy the so-called rules. The females are larger and more brilliantly colored or patterned than the males. They acquire the territories, aggressively do the courting, and defend their mates and nests from predators. In turn, the drabber males build the nests, incubate the eggs, and rear the young. In most polyandrous species, the female has little to do with the brood once the eggs have been laid.

In a polyandrous mating system, the female mates with more than one male, generally laying a clutch of eggs for one male and then going off to do the same for another.

The Spotted Sandpiper of North America generally is a polyandrous species. (In poor years or in areas where the habitat is sparse, Spotted Sandpipers become monogamous.) During the breeding season, the females fight each other for breeding territories and feeding territories, as well as for male birds. Each female is capable of laying up to four clutches of eggs in succession, making it important that she find as many males as possible.

PROMISCUOUS BIRDS

If birds had the avian equivalent of a singles bar, it would undoubtedly be the leks or arenas where male birds gather to parade their physical attributes in front of the females who come to have their eggs fertilized.

With no encumbrances, no ties that bind for longer than the act of copulation, these birds come together, mate, and then go their separate ways – the male to perform his display for yet another female, the female to lay her eggs and eventually raise the chicks that hatch.

A conservatively estimated 6 percent of all birds are promiscuous. In this sexual system, the male bird's role in reproduction is nothing more than that of fertilization. The only real effort he must make is in competing with the other males. This competition, however, cannot be underestimated. It is crucial because such a mating system usually ensures that only a few dominant males account for the great majority of successful copulations.

Birds practicing sexual promiscuity include the Ruff, many species of grouse, some pheasants, cotingas, manakins, many birds-of-paradise, many hummingbirds, most honeyguides, and bowerbirds.

These sexually promiscuous males differ – usually markedly so – from the females of the species. The males may have elaborate plumage, be brilliantly colored, make distinctive calls, or all three. Since the pairing period is brief, these distinctive features help to ensure that females choose males of their own kind and not of some related species. Thus, the chance of successful breeding is not wasted.

Promiscuous male birds may gather in leks to display themselves. The Sage Grouse begins displaying in early spring for three or four hours a day. With its tail cocked, the white feathers of the neck raised in a ruff, the male grouse inflates an air sac in its throat, contracts the throat muscles, and emits a loud "pop" as it expels the air. Much of this display is designed for the other males, since hens visit the lek only for a limited number of days in a season. On her one visit to the lek, a hen will make her way to the center of the lek, the domain of the more experienced males, ignoring the males on the periphery. The vast majority of all copulations are performed by one or two senior birds, with most males getting little for their efforts.

Ornithologists have long wondered why the males gather when only a few benefit. It was once believed that a decreased risk of being caught off guard by a predator might be one reason, but this conspicuous display may, in fact, attract more predators that it dissuades. Two sounder hypotheses are that males gather at sites where they are most likely to find roaming females, and that males gather because females prefer to choose fitter mates from among large groups of males.

As for the young males on the periphery of the lek, eventually some of them do have their day. Young birds grow older and aging birds die. Gradually, the males move toward the center of the lek, to be replaced on the outskirts by a new crop of eager young males who must dominate the sidelines and pay their dues for a few seasons.

Recent research indicates that some males are successful in breeding with females that travel from lek to lek. These males wait for the females out of vision of the lek males. Females may have to copulate a number of times in order to fertilize all of their eggs, and their timing may be such that matings away from the lek occasionally are essential.

COOPERATIVE BREEDING

In a system that, on the surface at least, appears to defy all evolutionary sense, some members of at least three hundred species of birds forgo their own reproductive opportunities and instead spend their time helping to care for the offspring of others. In such species it is common for birds of one or two years of age and even occasionally older to act as "helpers" for one or more years. In addition to these young birds, in some species adults who have lost a mate may also join in the efforts of another pair.

Cooperative breeding is most common in tropical and subtropical species that nest in colonies or have permanent pairs that occupy year-round territories. These helpers at the nest may share nest-building, incubation, feeding of the young, and territorial behavior with the breeding pair. Or they may serve mainly to bring food to the nestlings and help protect the young from predators.

The degree to which cooperative breeding occurs in a species varies. West Indian Todies and European Long-tailed Tits often have no more than one helping bird at a nest, while the Gray-breasted Jay (also called Ultramarine or Mexican Jay) and the White-winged Chough of Australia typically have several nonbreeding birds for each nestful of young birds. It is important to note that not all pairs in these species have helpers – this, too, can vary from year to year.

In another form of cooperation called communal breeding, several breeding pairs join together. The Anis, members of the cuckoo family, live communally in groups of up to four pairs. The females all lay in one large nest and take turns incubating the eggs. The glitch in this system is that there can be too many eggs to be incubated properly. If so, a female will roll existing eggs out of the nest before laying her own. The dominant female may lay last, which means that her eggs are more apt to stay in the nest.

The most obvious question when it comes to cooperative breeding is why birds do it. Are these cooperative breeders

A pair of Common Terns in the copulation position, female below, wings quivering.

prime examples of avian altruism or are they simply opportunists waiting for the right chance to strike out on their own? While appearances may make a case for the altruistic side of cooperative breeding, evidence points to the opportunistic advantages.

Normally, cooperative breeding has evolved in species where there is a dearth of optimum breeding territory, which limits breeding opportunities for the majority of adults, especially younger, less experienced ones. By helping to rear another bird's brood, the nonbreeders may enhance their chances of inheriting the territory. In some species, helpers actually recruit younger birds to assist them in overthrowing the breeding pair. In most cases, however, the pair simply do not drive away their young – the majority of helpers in most cooperative breeding species are related to the breeders.

Then, too, since birds with helpers are often able to successfully raise more nestlings than those without, the territories keep expanding as the size of the group grows. In some species, when the territory is large enough, one section is taken by the oldest male helper, who then pairs with a female from outside the family unit.

In most such species, the helpers are the progeny, mainly of one sex, of the breeding pair. Thus, by helping raise siblings, the bird is furthering its genetic heritage. Yet as a helper, a Gray-crowned Babbler rears an average of 0.46 fledglings, compared to the 3.62 fledglings it could raise if it were to breed. So why doesn't the young male find a mate and shed the parental yoke? One reason may be that the bird isn't yet capable of successfully nesting on its own. Another may be that the parents have an arsenal of tricks to keep their servant in servitude. In any case, the majority of young that stay to help parents are males – it is the young females who leave, thus preventing inbreeding.

Most humans, if hard pressed, don't mind indulging in a little in-law bashing. But if you think you've got troubles with your spouse's family, they're probably minor league when compared to the White-throated Bee-eater.

When this brilliant bird, with its emerald back, blue tail, and shimmering patches of red and black, starts attempting to nest, the young female moves into her mate's territory, ready and willing to lay eggs and raise a brood of little bee-eaters.

Then comes her father-in-law. His mission? To lure his son back to the family nest, where he will be expected to help his parents feed and care for the next brood of chicks.

The senior bee-eater is no dummy. He doesn't try to bully his son or daughter-in-law. The size of a thrush, the bee-eater is in no position to rely on brute strength. So instead he attempts to pester the pair.

The male's father becomes an almost permanent fixture at the young pair's nest. Quivering his tail, chattering his bill, and chirruping – all the friendly gestures a bee-eater makes in social settings – the father, nevertheless, interferes in the lives of the younger birds. Dropping in dozens of times a day, he does everything he can to disrupt the pair's efforts at nest-building. He parks himself at the entrance to the nest, making it difficult for them to come and go. When the younger male brings food to help prepare the female for egg-laying, the father begs for a taste and ends up with the lion's share.

It isn't hard to guess the toll this interference takes on the new pair. In about 40 percent of cases, the young male deserts his mate and dutifully follows Dad back to the nest to begin the task of helping to raise his younger siblings.

As for the deserted female, she finds herself in her newly appointed nest with little to keep her occupied. Sometimes she has already laid some eggs, but no matter how hard she tries, without the assistance of her mate, the offspring cannot survive. Meanwhile, in the nearby nest of her former in-laws, their new brood is flourishing with the help of her former mate. Alas, she too may end up helping some other pair to rear its young, unless another young male loses his father to a predator, and is freed to mate.

The Reproductive System

For most of the year, the reproductive organs of a European Starling are almost invisible, making flight an easier task. But as breeding season approaches, these tiny organs begin to grow until they are fifteen hundred times the normal size.

The starling is by no means unusual. In the bird world, nature has found a way to deal with what would just be excess baggage most of the year: get rid of it. Then bring it back when the need is there.

It isn't difficult to understand that even the very smallest organ enlarged hundreds of times can have a major impact on the body of a small creature. The testes of a drake, for example, may account for one-tenth of its body weight during breeding season.

The primary organs of the male bird's reproductive system are the bean-shaped testes, responsible for the production of sperm and the secretion of hormones. Mature sperm move out of the testes and into the vas deferens, a wavy tubelike structure that functions as a sperm holding tank.

Sperm cells do not develop in high temperatures. In man and other warm-blooded mammals, the testes hang in a scrotal sac outside the body, keeping them at the lower temperature that is favorable for their function. A similar arrangement in birds would have a disastrous effect on the animal's aerodynamic efficiency. So nature has devised a couple of ways to house the testes internally, yet keep the sperm cool.

In some birds, sperm production occurs only at night when body temperature is down. In other species, the storage section of the vas deferens is found in a cloacal protuberance where the temperature is several degrees cooler than the internal body temperature.

The female's equivalent to the male testes is the ovary. Most birds develop only one ovary, the left; the right ovary and oviduct appear early in development, then atrophy and disappear. During the breeding season, the ovary resembles a cluster of grapes. Eventually, certain "grapes" will become the yolk of an egg.

Most birds do not have external genitalia. Exceptions are some large species such as ratites, tinamous, pheasants, grouse, quails, and their relatives, curassows, storks, flamingos, ducks, and geese. A male of any one of these species has a form of a penis, which he uses to guide the sperm into the female's cloaca.

In most other birds, copulation usually involves ten or more seconds of contact between the male and female cloacas. Standing or treading on the female's back, the male twists his tail under hers, while she moves into a receptive position. Some birds such as swifts and swallows mate in midair. The male's position, whether it be in the air, on the land, or in the water, is often a bit shaky, a factor that may contribute to the brevity of the act, which is usually repeated frequently when both sexes are ready to mate.

A single ejaculation contains 1.7 to 3.5 billion spermatozoa. Fertilization generally occurs within a few days of copulation.

In species in which the male and female stay together at least until the eggs are laid, copulation is frequent, with many pairs mating ten, twenty, or more times a day prior to egg-laying. Copulation, even when it does not result in fertilized eggs, may serve the purposes of maintaining the pair bond and synchronizing the breeding physiology of the pair.

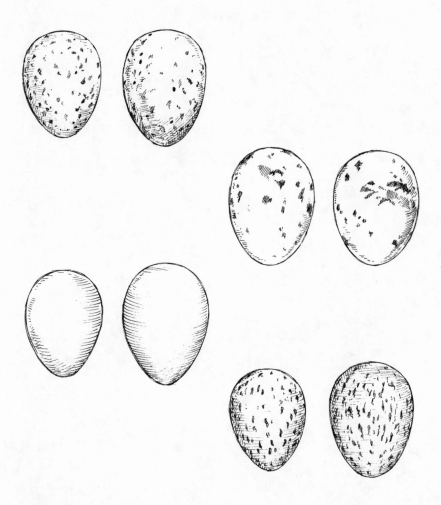

The eggs of four bird species, each paired with the type of Eurasian Cuckoo egg most often found in its nest. The eggs show a remarkable similarity, attesting to the evolutionary adaptations the Eurasian Cuckoo has made in its quest to survive. From left to right, in turn, the pairs of eggs are those of the Garden Warbler, cuckoo; Great Reed Warbler, cuckoo; Eurasian Redstart, cuckoo; White (or Pied) Wagtail, cuckoo.

10

Egg-Laying and Incubation

The Brown Booby, a widespread coastal and oceanic seabird that nests on islands, typically lays two eggs. Yet in most cases only one chick is raised.

Unlike many species that lay more than one egg, the boobies begin incubating immediately after the female lays the first egg. Then, often as many as seven days later, her second egg is laid.

Thus, the first chick is bound to hatch days before its sibling. As a result, this first chick is destined to monopolize its parents' time and energy. If the second egg does hatch, this weak chick will be hard-pressed to compete with its stronger sibling for food. Moreover, it endures outright abuse from its larger brother or sister, bullying that at its most extreme may take the form of cannibalism, a relatively common occurrence in the nests of pelicans, storks, herons, eagles, hawks, terns, owls, and any other species in which the hatching of the chicks is staggered significantly.

As a result, fewer than 1 percent of Brown Booby nests in a large colony surveyed in the Bahamas contained two live chicks, a percentage believed to reflect the norm throughout the species.

Why would a bird go to the trouble of producing an egg, incubate it, and then let it die of neglect or cannibalism?

Some ornithologists speculate that this may be a form of birth control, albeit a primitive one. The bird knows it has the resources to support only one chick, so it does nothing to enhance the chances that the other will survive. The obvious question is why bother with the second egg at all? The most likely explanation is that the second egg is the bird's "insurance policy." In any bird's nest, there may be an egg that fails to hatch for whatever reason. Thus, the booby insures its chances of producing one live chick by the female laying two eggs.

The majority of species are not as wasteful as the Brown Booby. Eons of natural selection have resulted in adjustment of the size of each individual species' clutch (the eggs in a nest) so that a bird lays only as many eggs as it can care for. Most birds also wait to incubate the eggs until the clutch is complete, insuring that the eggs will hatch at approximately the same time, and typically they do within a 24-hour period.

LAYING THE EGG

A female in the act of laying an egg typically rests on her heels, with body upright and feathers fluffed. She begins to strain or bear down as her abdominal muscles contract every few seconds. With each contraction, the cloacal opening widens until it is large enough for the egg to pass through.

Like its reptilian distant relatives, the bird lays an egg with a tough, water-resistant outer membrane. Moreover, the egg itself is a food store for the developing chick. Unlike the reptilian egg, the bird's egg has a hard, protective outer shell, which nevertheless is porous enough to allow air to penetrate to the embryo.

Two features, however, the standard of caring for the egg and the relationship the parent bird has with its young, broaden the gap between the worlds of the bird and the reptile. The female reptile generally lays her eggs in or on the ground, whereas most birds build some type of protective nest. The majority of reptiles abandon their eggs to fate. Most birds, on the other hand, warm the eggs with the heat from their own bodies until the eggs develop enough to hatch. Even then, for many birds the task is far from complete. The nestlings are warmed (or kept shaded and cool) and fed, protected, and nurtured by one or both parents after they leave the nest.

The laying of a bird's eggs exacts a steep price from the mother bird. A female chicken deposits a daily sum of 1.8 percent of her overall body weight in her developing egg. Compare that to the 0.019 percent of body weight that the human female loses to her developing fetus each day.

Interestingly, the process of laying the egg itself is usually brief. The Brown-headed Cowbird (a brood parasite under pressure to lay eggs rapidly) releases an egg in a few seconds, while a Bobwhite Quail requires anywhere from one to three minutes to lay her egg. Then there are the birds that must work harder to release their eggs. Larger birds such as Turkeys and geese have been observed straining for one or two hours before the egg drops out.

Some birds have little control over the moment when an egg is laid. If you happened to be holding a Prairie Warbler at the time her egg was ready to be released, you would end up with it in your hand. Other species have precise control over the moment when the egg is dropped. A cuckoo, which lays her eggs in the nests of other species, can hold in her egg until the nest owner leaves for a few moments. Then she flies to the nest, surreptitiously deposits the egg in a matter of seconds, and flies away, never to return, although she may spend a few moments carrying away one of the host's eggs with her to help "fool" the other bird.

A bird is capable of laying only one egg at a time. Most of the birds we see at our backyard bird feeders (as well as ducks, geese, chickens, and shorebirds) can lay an egg every day until their clutch is completed. At the other extreme are the megapodes – birds that lay their eggs in the ground and let volcanic heat, the heat of the sun, or heat from decomposing vegetation keep them warm – that require as many as eight days between the production of their unusually large eggs. The interval between the laying of each two eggs probably corresponds to the time it takes to form the egg's many layers.

The time of year in which egg-laying takes place ensures that the young are scheduled to hatch when conditions are most favorable to their survival. Sometimes this may require laying the eggs in bad weather, if the incubation period is especially long. The Great Horned Owls of Pennsylvania and Iowa must lay their eggs in February, one of the coldest months of the year. Yet by the time the young owls hatch in spring, an abundant crop of rodents is theirs for the parents' taking. In contrast, the Eleonora's Falcons of the Mediterranean islands lay eggs in the summer, so that the young falcons are raised on easily caught small birds that have begun their fall migration.

In many species, the approach of spring is closely synchronized with egg-laying. Nesting records of the British Trust for Ornithology have shown that, at sea level, the Meadow Pipit lays its eggs on the average 3.8 days earlier in the south of Britain than in the colder north. As the elevation increases, so does the delay in egg-laying.

Some species will lay their eggs earlier in years when spring weather is good. European Alpine Swifts, according to one study, began nesting on May 17 for nine years running when the spring weather was dry and warm. During eight springs of cooler, wet weather, however, the swifts postponed nesting until May 31. Naturally, not all pairs begin laying exactly at the same time – early nesters may be favored one year and later nesters the next.

Many birds rely on specific cues to commence reproduction. Great Crested Grebes usually will not breed until the reeds around their floating nests have grown to a certain height, and Mew Gulls and Herring Gulls in the Arctic also postpone breeding on islets until the surrounding ice has melted, which prevents the Arctic fox from stealing their eggs.

The age of a bird also influences when its eggs are laid. Females of some species can breed when they are only weeks old. A captive Common Quail may begin laying eggs when she is only thirty-eight days old. Domestic hens generally lay their first clutch of eggs when they are five to seven months old, while most passerines, pigeons, and ducks lay at one year. At the other extreme is the female Royal Albatross, which does not begin to lay eggs until she is eight years old. As a general rule, a novice female will lay her eggs a few days later than one who has had prior experience.

While poor weather, inadequate food, or lack of opportunity may delay some species' laying of eggs, some birds are remarkably consistent whatever the conditions. After a winter in the north Pacific, the Short-tailed Shearwater returns to its breeding grounds on the islands of the Bass Strait in Australia. Year after year, the majority of these birds lay their single egg on or within two days of November 25.

Many birds are also consistent when it comes to the time of day they lay their eggs. The early morning hours, even before dawn, are a common time. Some speculate that this may be because an ideal time for eggshell formation is during the night when the bird is inactive. The bird greets the morning with a heavy and delicate egg in its oviduct, which from a physical standpoint isn't conducive to doing much of anything except laying it.

Of course, not all birds lay their eggs early in the day, nor do they lay each one at the same time of day. The only thing consistent about the Ringed Plover's laying schedule is its inconsistency; the American Coot appears to be partial to laying just after midnight; then there are birds such as the Redwing, a thrush that breeds in Swedish Lapland. This bird lays one of its five or six eggs every twenty hours, meaning that, before its clutch is complete, it has laid an egg during most times of the day. The rapid completion of the clutch allows for maximum use of the short northern summer, the only period when there is abundant food for the young.

EGG SIZE AND COLOR

The Common Murre nests in large colonies, each female laying her single egg on an exposed rock ledge. If all the adults are flushed from the colony by the approach of humans, they leave a colorful sight behind. One large egg on one ledge is a deep blue-green. Another ledge contains a bright reddish-colored egg. Across the ledge sits a creamy white egg, and around the corner is a blue one. One white egg on the far side of the colony has lacelike patterns of yellow and black, the one next to it is as unblemished as a young child's skin, while the one nearest is blotched with red, uncannily like a teenager's complexion.

The Common Murre, though, is an exceptional case. Usually the eggs of a particular species of birds are about the same color and more or less uniform enough in size and shape so that anyone can readily identify them. It is believed that the great variation in color and pattern in the Common Murre may enable the individual murre to identify its own egg from so many others in the colony. Numerous experiments with this species have shown that murres will accept a strange egg if the color and pattern are similar to its own. When it is noticeably different, the bird readily rejects it.

In the beginning, all birds' eggs were probably white. Natural selection most likely favored the evolution of egg color, probably as a way of camouflaging eggs against predators. Supporting this theory is the fact that most modern birds that still lay white eggs are hole-nesters (some swifts and owls, parrots, woodpeckers, kingfishers, and bee-eaters), birds nesting in open nests but starting incubation immediately after the first egg is laid (some doves, herons, hummingbirds, owls, and grebes), or birds that cover their eggs to hide them when they must leave the nest (some ducks, geese, and grebes). The majority of other birds laying in open nests produce eggs that blend with the surroundings.

For most ground-nesting species, this means the egg is a brownish color if the bird is laying on dead vegetation, a light, sandy color if it nests on beaches or deserts, and even black if it builds its nest on ground that has been burned. The family of thrushes is ones whose members' nests vary according to nesting site. Ground-nesting thrushes produce brown, gray, or olive-colored eggs. Those building their nests in the fork of a tree produce heavily blotched eggs, while crevice-nesters have speckled white or blue or just plain blue eggs. The thrushes that lay their eggs in holes produce white eggs.

One bird especially adept at matching its eggs to the ground on which they are laid is the Yellow-wattled Lapwing, a bird of India. One coastal region in which the bird lays has a brick-red, sandy soil through which black particles of ironstone are scattered. Only in this area does the lapwing lay red eggs covered with brownish specks.

Ornithologists have experimented with egg color to determine whether this is indeed a factor in protecting the developing chick within from its enemies. In one experiment on the edges of rookeries of Common Black-headed Gulls, scientists deposited an equal number of simulated naturally spotted gull eggs, white eggs, and khaki-colored eggs. Herring Gulls and Carrion Crows, two of the Common Black-headed Gull's predators, were more likely to take the plain-colored eggs than the naturally spotted ones.

Like color, egg size in the bird world also varies greatly. The smallest egg is laid by a hummingbird and weighs about 0.02 ounce. The largest is that of an Ostrich, weighing on average 3.3 pounds. There is individual variation in egg size, and birds with a large clutch size usually lay slightly smaller eggs as the clutch is completed, the earlier eggs being a bit larger.

The relative size range of bird eggs: from left to right, Ostrich, domestic hen, and Ruby-throated Hummingbird.

As a rule, large birds lay large eggs, small birds small eggs. This can be misleading, however, because small birds tend to lay much larger eggs relative to their body mass than do most larger birds.

The Ostrich, the largest bird in the world, lays the largest egg. Nevertheless, that egg weighs only 1.7 percent of its body weight. Compare that to the modest-sized kiwis, which probably hold the world's record for laying the largest egg in relation to their size. The chicken-sized female produces an egg equivalent to 25 percent of her body weight.

The developmental stage at which the chick is hatched also plays a role in egg size. Typically, precocial species are more likely to come from larger eggs than the naked and virtually helpless altricial young. Although the guillemot and the raven are birds of equal size, the egg of the precocial guillemot is five times larger than that of the altricial raven. This reflects the need for much more complete development of precocial young inside the egg – in altricial birds such development is postponed until after hatching.

Even though the basic size of a species' egg is inherited, some variables have an influence over the weight of an individual bird's eggs. Thus, a larger female bird ordinarily produces larger eggs than a smaller female living on an adjacent territory. A female breeding for the first time also may produce smaller eggs than her more experienced counterparts, while a very old female's eggs also may be smaller than those from females laying at the prime of their lives. Another influencing factor is the amount of food to which an egg-producing female has access. A bird that begins to lay her eggs in the midst of a food shortage undoubtedly will produce smaller eggs than she would if food were plentiful.

Within a single species, egg size may differ in different geographical areas. Widespread species show size variation; the more southern birds from nearer the tropics are smaller than those from farther north (and vice versa in the Southern Hemisphere). Egg size shows a similar variation.

Eggs laid in the same clutch may differ in size. Although the domestic hen's first egg is slightly smaller than the rest, for other birds it is more often the last egg that is the runt of the nest.

The explanation for unusually small eggs is generally thought to be the female's lessened energy. In Western House Martins, however, small eggs are more likely to be laid when bad weather causes the supply of flying insects to dwindle. Since most of the production of an egg occurs within a very few days, the female's intake of food obviously will have some effect on egg size.

Regardless of the cause, the future of these runt eggs is often abbreviated. First of all, they may not hatch. If they do, the resulting chicks are usually smaller, and if food is scarce, they are the first to die because they are unable to compete for food. In some owls, the later eggs in the clutch decrease in size with laying order. One by one, the smaller nestlings die until the size of the brood is compatible with the existing food supply. Only in very good times will the smallest bird of the nest live to leave home.

CLUTCH SIZE

In 1944, what is believed to be the egg-laying record of a European Robin was set when an observer watched one bird lay twenty eggs in a nest built to comfortably house a fraction of that number. By the time the bird had deposited the twentieth egg into the cup-shaped nest of moss and leaves, the eggs were piled three deep. The interested human, watching the bird hard at work from a distance, was convinced she would have continued had she not been forced to desert her nest and its contents by the ill-timed arrival of a cat.

Typically, a European Robin lays a clutch of five eggs. Some individuals lay one less; others, one or two more. Occasionally, an ornithologist will stumble upon a nest containing as many as twelve eggs. When this occurs, some of the eggs are often abnormal, with many failing to hatch.

The number of eggs any bird lays – its clutch – is, like so many elements of a bird's life, determined by natural selection. At least three main factors seem to have affected the evolution of the average size of a given species' clutch, although there is room for individual and even geographic variation.

First, the more eggs in a clutch, the harder the parents must work to raise the nestlings. The parents of an average-sized brood may work most of the day to keep them fed; a larger clutch doesn't necessarily mean they will step up their efforts because they may be already working at their maximum capacity. In general, then, each individual from a smaller clutch is likely to be better fed than those hatched in a larger family. Thus, the nestlings that hatch from a larger clutch of eggs, if they survive, tend to leave the nest weighing less than other birds, a disadvantage that equates with a lower rate of survival.

The second factor that probably regulates clutch size is how big it can be and still remain relatively inconspicuous to predators.

Lastly, there's the survival of the parent birds. The larger the brood, the more "wear and tear" on the parents, which, after all, must have time to maintain themselves – to eat, preen, and take sufficient care of themselves to sustain the nesting effort. From a cost-benefit standpoint, a successful bird is one that over a lifetime leaves more than one bearer of its genes to replace it. To the extent that an overly large clutch wastes effort and risks the lives of the pair, natural selection has honed the number in the clutch to an optimal size.

Thus, among the many species inhabiting the bird world, the variation in clutch size is truly astonishing. Like the endangered California Condor, birds such as shearwaters, petrels, penguins, and albatrosses bestow their efforts on a single egg. Kiwis, loons, boobies, many eagles, most pigeons and doves, and most hummingbirds produce two eggs. Most gulls and terns lay three eggs, while the nest of a plover or sandpiper usually contains four eggs. Passerine species living at higher altitudes generally lay four to six eggs. Some titmice and ducks are among the most prolific, with nests that usually contain from eight to twelve eggs.

Within the given parameters of a species' genetics, many environmental and individual factors can influence clutch size. The food supply is a vital one. By the time some species of tits have finished laying, the weight of the clutch of eggs may exceed that of the mother bird. The production of such a clutch requires large amounts of extra food. If that food is not available, the size of the clutch will shrink.

Another bird whose clutch size reflects its fat stores is the Lesser Snow Goose, which breeds in the highest reaches of the Arctic, arriving when the ground is still blanketed with snow. The arriving geese, already mated, lay their eggs and sometimes begin incubating before the climate enables them to forage for food. During these lean days, the mother must rely on her fat reserves. If she is well fed, her clutch will be larger than if she has little extra stored fat.

Clutch size may also be determined by feeding methods. Birds like Whippoorwills, that can forage all night but favor feeding during the brief twilight hours of morning and evening, usually lay only two eggs, while large seabirds that commonly travel hundreds of miles a day to find food for their young typically lay only one egg. This is not simply a case of a bird being overly conservative. For many of these long-distance foragers, raising one chick is a monumental effort. Consider the Red-footed Booby of the Galapagos Islands. Studies have shown that nearly 70 percent of the single eggs laid by this bird end up being abandoned before they hatch because one long-distance foraging parent, due to the lack of nearby food, is unable to return to the nest on schedule to replace its incubating mate.

The size of a bird's nest often bears some relationship to its clutch size. To see whether nest size would indeed make a difference to the number of eggs a Great Tit laid, researchers dotted a German forest with two sizes of nest boxes, one more than twice the size of the other. During a two-year period, the Great Tits of the forest laid and hatched more eggs per nest in the larger boxes. In the midst of the experiment, the size of the boxes was changed, and they were replaced with boxes of other sizes. Again, the birds laid more eggs in the larger boxes.

The latitude at which a bird breeds and nests appears also to make a difference to the clutch size. Generally, birds breeding in temperate areas north and south of the tropics lay more eggs than the tropical breeders. This holds true even in birds belonging to the same species.

This difference is largely attributed to the effect of the longer days of the northern latitudes, which allow more time to find food, thus making it possible to support a larger family. A related factor is that the time available for nesting in the far north is too short to permit the raising of more than one brood in a year. And there are adaptations to permit larger clutches in some species. The Snowy Owl of the far north lays more eggs than do related owls who live in more southerly climates. To do this it must feed throughout the very long summer days in the north, during which the sky is never totally dark.

One credible theory is that the size of a bird's clutch is related to the abundance of food per bird. Unlike the far north and south, where food is plentiful only during a limited period, food supplies in much of the tropics are stable throughout the year. In addition, the amount of daylight is fairly constant during the year, allowing more hours of feeding per day. Breeding in the tropics tends to be spread over more of the year, reducing competition for food at any one time and also permitting several broods a year. As a result, smaller broods are favored in warmer climates. As for the northern breeders, they must literally put "all (or more of) their eggs in one basket."

As a result, within geographically widespread species there may be a geographic variation of clutch size. The European Robin that lays an average of 3.5 eggs in the Canary Islands would lay 4.9 eggs in Spain, 5.8 eggs in Holland, and 6.3 in the far northern outpost of the species in Finland.

An interesting geographic variation occurs between some island-nesting species and their mainland counterparts. Coal Tits on the island of Corsica lay clutches averaging two to five eggs less than Coal Tits that reside on the French mainland. Similarly, several species of birds that live in Britain lay fewer eggs than their European counterparts. The reason for this difference? No one knows for certain, but islands do generally support fewer species with less competition per species, often allowing higher population densities. As the density of a breeding population increases, the size of the clutch often decreases.

MANIPULATING A BIRD'S NUMBER OF EGGS

The California Condor normally lays one egg on a bed of coarse gravel on a cliff, cave floor, or in a cavity in the trunk of a giant sequoia. Unlike most species, the condor does not breed every year, although it is physiologically capable of doing so.

In recent years this condor has become yet another tragic victim of man's zeal to leave no stone unturned in his quest for supremacy over his environment. The California Condor could once be found throughout the southwestern United States and northern Mexico. Then hunters shot at the soaring birds for sport. Many died from eating poisoned food meant for coyotes; others succumbed to the effects of DDT and other pesticides in the food chain, and from lead poisoning in food and water supplies. Condor eggs were snatched from their nests by unscrupulous collectors. The food supply diminished for this bird, which thrives on dead carcasses of mammals and fish. And natural habitats were replaced by ranches, suburban houses, and huge estates.

By World War II, the total condor population was estimated at slightly more than sixty birds inhabiting hundreds of miles of California's mountain ranges. By the late 1980s, not one wild condor remained in the mountains of California; the remaining twenty condors were alive in captivity.

The conservation community had long pondered the issue of how to save the condors. A controversial program was ultimately adopted that involved captive breeding.

Ornithologists working in aviaries where the captive condors were housed observed that if the bird's one egg was removed from its nesting spot, the mother simply laid another one. When the second egg was removed immediately after laying, she was likely to replace it with a third egg. Thus, instead of adding one chick to the condor population, a female could produce as many as three if the situation was manipulated properly. In essence, then, the world would gain three nestling condors instead of one. According to some, this strategy would allow the fate of the dying breed to be turned around.

Detractors, however, question whether the laboratory-bred birds would bear any resemblance to their proud and free ancestors. Of most concern is the problem of what the lack of natural selection means to the genetics of the condor population and its ability to adapt to the changed conditions in prospective habitats in the wild.

Only time will tell whether this ongoing experiment with captive breeding will ultimately enable the condor to live and breed in the wild. Several of the aviary-bred and raised condors have now been put back in the wild.

Condors are not the only birds that lay an egg to replace one that disappears. Many species are "indeterminate layers" – that is, they are capable of laying as many eggs as it takes to complete the clutch. Among species that lay eggs in such a manner are some penguins, ducks, chickens, some woodpeckers, and some passerines.

In one famous experiment with a Northern Flicker, scientists removed the bird's eggs one at a time from her nest as quickly as she laid them. Under normal circumstances, the Northern Flicker lays six to eight eggs. In this case, however, this female laid seventy-one eggs in seventy-three days before, physiologically if not physically exhausted, she quit laying.

Nowhere is manipulation more apparent than in the laying practices of domestic fowls whose eggs are eaten by humans and constitute a multi-million-dollar industry. One Japanese Quail was induced to lay 365 eggs in one year, while some domestic hens (with the best food and care, of course) are capable of laying 352 eggs in 359 days. These, of course, are extremes coaxed by the bottom line on a business ledger. In the wild, an indeterminate layer automatically stops laying when her clutch is complete.

Not all birds are capable of replacing a lost egg. Petrels seem capable of laying but one egg per season; if the egg is destroyed, the petrel has no choice but to wait until the next year.

So fixed is the number of eggs in some species that no manipulation will cause the bird to deviate. Remove an egg from the nest of a crow, and the bird will incubate the remaining eggs despite the less-than-normal number in the nest. Or, if an additional egg is added to the nest, the female still will lay her usual number and attempt to incubate the larger-than-usual clutch. These so-called determinate layers respond neither to missing nor additional eggs in the nest, but appear to be programmed to lay a certain number regardless of outside stimuli. Of course, in the wild this manipulation simply doesn't happen. Any predation usually involves not one but all the eggs and if the nest is damaged, the bird abandons it.

Some birds such as the Herring Gull will replace their eggs only if the entire clutch is lost, but they do not replace the eggs in the same nest. Rather, the pair desert the nest, build another one elsewhere, and begin laying again.

Regarding indeterminate layers, the obvious question is how do they know when they have laid the proper number of eggs? Studies suggest that the laying female is able to feel the correct number of eggs against her abdomen, which somehow acts as a signal triggering the endocrine glands to halt egg production.

The way in which this occurs was studied in the Common Black-headed Gull. During courtship, the gull's hormone production increases, initiating a sequential development of the egg follicles in the ovary. Unlike many species, the female gull begins incubating after the first egg is laid, before the clutch is complete. The act of incubating physiologically inhibits further egg development, but does not halt the growth of the appropriate number already developed to a certain point. As a result, the egg that is most developed is laid, and often the one after that. But the fourth egg never reaches the point of adequate development and is reabsorbed by the mother's body.

INCUBATION

Winter temperatures in Antarctica drop as low as minus 77 degrees Fahrenheit – its climate is not one that easily sustains life. Yet at this moment, the Emperor Penguin is alive and very well in its Antarctic home.

When all the marvels of the avian world are considered – and there are so many – one that has to be described in more detail is the incubation feat performed by the Emperor Penguin. In its sub-zero surroundings, this penguin is able to maintain the temperature within its single egg at 93 degrees Fahrenheit. After the female lays her single egg, she leaves and the male immediately sinks atop it, with the egg on his feet. Thus held, on its father's webbed feet, enveloped within a fold of belly skin, the egg sits for eight to nine weeks, its warmth a sharp contrast to the frigid air outside. The parent bird, unable to do anything except care for his charge, lives off his enormous fat stores until his mate returns from feeding in the sea, about the time the egg hatches.

The main task of the incubating bird is to transfer heat from its body to the developing chick within the egg. There are several ways in which birds do this. Cormorants, gannets, and frigatebirds wrap their webbed feet around their eggs to provide warmth. The so-called incubator bird, the Egyptian-plover, buries its eggs in hot, sun-drenched sand. When the eggs become overly hot, the adult may act as an awning, standing over the eggs, or it may wet its plumage and drip water on the sand over the eggs.

The scrub-fowl prefers to use warm sand or soil heated by volcanic steam. But when other scrub-fowls live in the forest and these resources are unavailable, they rake fallen leaves into a pile of appropriate size (about thirty-three feet across and sixteen feet high) and lay their eggs, covering them in the pile. The eggs are then incubated by the heat of the fermenting vegetation. The male checks the temperature around the eggs frequently, taking off vegetation to cool them or adding more to hold or increase the temperature.

While these are interesting and innovative methods of incubation, by far the most common way in which eggs are incubated is by the parent sitting on them. The period of time a bird must sit on its eggs until they hatch varies. One factor in determining incubation time is the weather. In warmer seasons, the incubation period tends to be shorter, while in cooler weather the bird must spend more of its time on the nest.

As a result of temperature fluctuations and the varying needs of the parents to feed and sustain themselves, the incubation period even within a species will vary significantly. Heron eggs are incubated anywhere from seventeen to twenty-eight days, while in hawks the range is from twenty-eight to thirty-eight days.

Although there are exceptions, usually the smaller the egg, the shorter the incubation time. Thus, some of the small passerine birds incubate for as little as eleven days, while the Royal Albatross sits on its larger egg for eighty days. Brood parasites that lay their eggs in the nests of other species tend to have very short incubation periods, ensuring that the young parasite hatches first.

Just because a bird is sitting on its nest, however, doesn't necessarily mean that it is incubating its eggs. A bird does not incubate with its feathers, which are, in fact, extremely poor heat conductors. As we saw in Chapter 2, many species of birds are equipped with a temporary blood-enriched area on the abdomen known as a brood patch. Just prior to incubation, the bird loses its feathers in the area of the patch and the skin becomes spongy, appearing inflamed from all the extra blood capillaries in the area. Some birds have a single patch, while in some species there may be several small patches, one for each egg. The brood patch may be located away from the abdomen in some birds. Boobies and gannets, which warm their eggs by standing on them, have heavily veined feet that provide additional warmth for the developing egg. On hot days or in the heat of the afternoon, the "incubating" bird may raise itself above the eggs, allowing the air to cool them down.

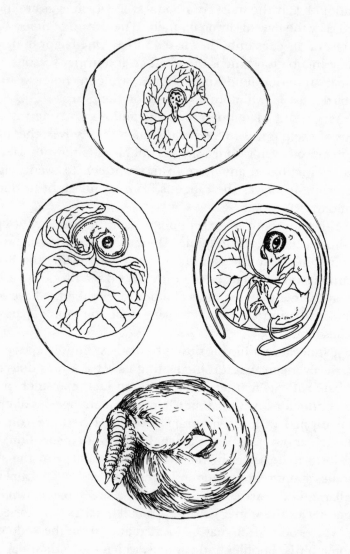

A generalized bird embryo (of, say, a medium-sized passerine), showing development from the early embryo up to roughly the nineteenth day after the onset of incubation.

Typically, it is only the parent responsible for incubating that develops the brood patch. Usually this is the female, although in some species it is the male that does the incubating and develops the patch. In species in which both partners incubate, both develop patches. In some passerine species, however, the male may sit in the nest to protect the eggs (e.g., while the female feeds) but does not have the ability to actually incubate them.

MAINTAINING THE TEMPERATURE

Few birds are as meticulous as the Malleefowl when it comes to testing temperature. This megapode male (scrub-fowl and Malleefowls are members of the megapode family) digs a pit about ten feet wide and three feet deep. He fills it with leaves and waits for the rains to begin the rotting process. When that is progressing well, he covers the pit with sand to contain the heat brought about by the decomposition. Over a period of several months, one or more females lay their eggs near the surface of the pile.

It is the male's job to keep careful watch over the temperature, maintaining it at 91 degrees Fahrenheit. By picking up a sample of the pit's vegetation in his beak he is able to determine the temperature precisely by a heat-sensitive internal "thermometer." If the pit is too hot, he opens up the pile of leaves to allow heat to escape. When the summer sun is too intense, the bird spreads more sand over the top to provide better insulation, allowing the leaves to cool. In the cooler fall, the fowl clears out sand in the morning so that the sun's warmth will penetrate to the eggs, and kicks it back on the pile in the evening to better insulate his progeny against the night chill.

An incubating bird does not "park" itself in a nest for a given number of days; it must not only warm the eggs but prevent them from becoming overheated. And it must adjust its actions to the climate and environmental conditions.

The temperature at which the eggs are kept is several degrees lower than that of the brood patch. As a result, the incubating parent must balance the heat transferred by its body with that of the day or night, as well as any heat generated by the developing embryo itself. If, for example, the day is warm, the eggs could get overheated if the adult bird covered them with its brood patch. So, instead, the bird may sit on the nest in such a way that a layer of its breast and flank feathers buffers the eggs and the heat-intensive brood patch.

Ornithologists have long experimented with various species to determine how a bird knows when an egg is too warm or cold. In one experiment, artificial copper eggs were pumped with temperature-controlled water. When the egg temperature was abnormally high, the incubating female Ring Dove extended her neck and opened her mouth in a flutter. When the eggs were cold, she shivered, fluffed out her feathers, and drew in her neck, actions both producing and retaining her body heat to enhance its effect through the incubation patch. This and similar experiments suggest the presence of sensory receptors in the brood patch itself.

For birds that lay their eggs in the north, the problem is mainly that of keeping the eggs warm, but for those in the tropics or that nest in the open on sand or stones, the greater task is keeping them cool. Many of these birds build domed nests or lay their eggs in protected cavities to add some much-needed protection from the brutality of the sun's rays. But the birds whose nests are exposed have had to come up with some innovative ways to protect their developing young. The Common Nighthawk sits on the ground close to its nest during the hottest part of the day, using its wings as a parasol over its eggs. Many desert birds such as the American White-winged Dove press their belly against their eggs to help cool them. The bird's body temperature, maintained lower than the external temperature, allows the dove to use the belly essentially as a sponge to absorb heat from the eggs.

The problem then becomes how to cool off the parent bird. The Indian Yellow-wattled Lapwing uses its wings like venetian blinds, ruffling the feathers so that they allow air to move, cooling the surface of the body, yet keeping the sun off the bird's skin. Feathers allow the bird to hold and maintain a layer of air against the skin that acts as insulation against heat loss. When the feathers are raised, this layer is broken and air reaches the skin, cooling it. Birds lucky enough to have a source of water nearby may wet their feathers or sprinkle water on the eggs themselves.

Generally, birds nesting in cooler climates spend more time incubating than those that lay eggs in warmer climates. In one study, nest boxes occupied by Pied Flycatchers were heated, the result being that the flycatchers spent less time incubating and more time feeding.

As might be expected, any time the incubating bird leaves the nest to forage for food, the eggs lose heat if they are not in the sun. The rate at which this occurs depends upon the size of the egg, its placement in the nest, the amount of nest insulation, and the external temperature. When the parent bird returns, of course, the eggs are reheated. Within a certain range, fluctuating temperatures are not dangerous to the developing chick. Moreover, in the latter half of its incubation, the chick's own increasing metabolism aids in slowing the rate of temperature cooling when the parent is away. At any rate, the nesting adults seem to be aware of the temperature because they always return at the correct time, unless the other parent has disappeared or food is in very short supply.

For some species, the most reliable indicator of egg temperature is the chick itself. Shortly before a young murre begins to break out of its shell, it makes several calls. One of these calls is made in distress when the egg is too cold. Hearing the call, the incubating parent quickly covers the egg more closely, which rapidly raises the temperature and soothes the chick.

INCUBATION PATTERNS

The changing of the guard over a nest of Brown Pelicans is something like this: The returning parent lands near the nest and slowly approaches, its bill held vertically and its head bobbing from side to side, then it pauses. Next, both the approaching bird and the incubating one begin to preen or groom their feathers. A moment later, the incubator vacates its spot and the newcomer steps over and begins its shift.

The ways in which male and female birds share (or don't share) their parental duties vary. By far the most common incubation pattern is for both the male and female to take turns. In one study of 160 species, 54 percent shared incubation.

This is not to say that the division of labor is always an equal one. Although in most species the male does spend some time on the nest, in many bird families it is the female who is the main brooder. In these species, the male's most important role is that of food gatherer.

Species whose males and females do roughly equal amounts of incubating include cormorants, gannets, herons, storks, auks, toucans, and woodpeckers. Petrels change shifts on the nest every seven hours, while female pigeons sit at night, and the males during the day. The Short-tailed Shearwater's shift on the nest is among the longest. The male sits on the nest for the first twelve to fourteen days while the females congregate in flocks to gather food at sea. At the end of her feeding period, the female takes over at the nest and it becomes the male's turn to begin foraging.

Both sexes of a few species even incubate simultaneously. The female European Red-legged Partridge builds two nests, filling the first with eggs. For fourteen days, the eggs remain unincubated while the female completes the task of filling the second nest. She then returns to the first nest and begins incubation, while her mate sits on the eggs in the second nest. Each parent cares for the young in its own nest.

Sometimes it makes sense that only one parent incubates the eggs. This often occurs in species in which one sex (usually the male) is so brilliantly colored that the bird's presence near the nest would increase the risk of a predator.

Just because the male does not incubate, does not mean he is idle during this period. In some hawks and owls, hornbills and others, it is the male who is responsible for keeping the female well fed during this time. The Red-billed Hornbill is an example of how draining a task this can be. The female hornbill imprisons herself in her tree-cavity nest by blocking all but a small slit in the entrance with mud, regurgitated food, and feces. When dried and hardened, this is a considerable fortress against predators. Through that slit, the male is able to pass whole and regurgitated fruit and insects to his mate. After her self-imposed imprisonment – which may last as long as four months – the female, fat and healthy, breaks through the wall of mud and emerges with the brood. The male, on the other hand, is thin and worn from the strain of supporting his family.

BROOD PARASITES

Not every species of bird assumes responsibility for its young, even if its hatchlings are altricial. When it is time to lay eggs, the Eurasian Cuckoo, for example, unlike most birds of the world, isn't concerned about building a nest in which to lay a clutch. Rather, females begin searching for suitable existing nests in which to lay. Then the female cuckoo waits until the owner of the nest is gone for a few moments. Quickly, she advances upon the nest, breaks or carries off one of the host's eggs, and within a few seconds deposits her own in its place. Then she flies away, not bothering to look back. Forty-eight hours later the process is repeated in yet another bird's nest. And so it goes, on and on, until the cuckoo has laid anywhere between sixteen and twenty-six eggs.

The Eurasian Cuckoo is but one – albeit the most famous – of the world's parasitic birds. An estimated 1 percent of all species lay their eggs in the nests of other birds. These so-called brood parasites include cuckoos, honeyguides, some cowbirds, some weavers, and the South American Black-headed Duck. In some parts of East Africa, as many as nineteen species of brood parasites may breed within a small area.

The origins of this somewhat strange behavior are unknown but attempts at explanation abound. One plausible theory is that the many physiological changes undergone during reproduction do not progress in a normal manner in these birds. Thus, parasitic species may be compelled to lay eggs before hormone levels awaken the instinct for nest building.

Other ornithologists have suggested that brood parasitism probably began in the tropics where the risk of nest predation is high. A female bird that loses her nest to a predator must quickly find another receptacle in which to lay her eggs, the most likely being the existing nest of another bird. Moreover, the loss of a nest actually triggers ovulation, causing her to lay even more eggs than she would under different circumstances.

Some have suggested that a precursor to brood parasitism may be the habit among some birds of using other species' abandoned nests for their eggs. Owls often nest in old crows' nests and starlings use woodpeckers' holes, sometimes even usurping them from the rightful owners.

Recent molecular studies of various species show that some individual females of many of these species actually lay one or more eggs in the nest of another pair of the same species. Such "dumping" probably has resulted in the evolution of brood parasitism in cowbirds and Black-headed Ducks.

However murky its beginnings, the advantages of parasitism are clear. By laying in many different nests, the bird enhances the chance that at least some offspring will survive. In addition, since it doesn't have to incubate the eggs and ultimately care for the nestlings, the brood parasite can lay a greater number of eggs.

It stands to reason that any bird laying its egg in another species' nest must be adapted somewhat to that particular species' lifestyle. Food requirements must be similar. The nest of the host must be easy to find and enter. Incubation and fledging periods must be roughly the same, and the parasite must be ready to lay its eggs at the same time as the host.

Luckily for them, most parasitic birds are amazingly adept at fitting in with their hosts.

Parasite parents don't simply choose another bird's nest at random; most are very particular about whose nests they will lay in. The Greater Honeyguide uses the nests of rollers, starlings, and bee-eaters, while the Scaly-throated Honeyguide prefers woodpeckers, and the Lesser Honeyguide looks for the nests of a tinkerbird or small barbet. These honeyguide hosts all nest in cavities or holes or have covered nests.

To increase the chances that its egg will fit in with those of its host, many nest parasites lay eggs similar in size and color. Old World cuckoos lay eggs that in color and markings mimic those of the host, with individual female cuckoos specializing in one species of host. The eggs of the Diederik Cuckoo, for example, are nearly identical to those of its host, the Vitelline Masked Weaver. The only way to differentiate between them is by an analysis of their structure or molecular features.

Usually a species' eggs vary in color. In Finland, the Eurasian Cuckoo lays blue eggs to match those of its hosts, the Eurasian Redstart and the Whinchat. But in Hungary, the same species of cuckoo lays greenish eggs with dark markings, similar to those of the Great Reed Warbler, to whose nests Hungarian cuckoos are most partial.

The ability to mimic closely the eggs of its host appears to be a factor in whether or not many host birds will accept the parasite's eggs. In a study of a cuckoo in India, host birds whose eggs matched those of the parasite deserted their nests only 8 percent of the time, compared with 24 percent of those hosts whose eggs were not similar.

Ultimately, of course, the fate of any parasitic species depends upon acceptance by the host. This varies greatly among species and sometimes within species. Some hosts occasionally will toss their own eggs out of the nest and care only for the parasite's egg; more often the host is apt to reject the new, perhaps odd egg. Some hosts of the Brown-headed Cowbird take no chances but rebuild a nest over the top of the old one, covering their own old eggs and laying a new clutch.

Can birds recognize their own eggs? Obviously, for a bird to throw out an egg similar to its own, it must be able to distinguish its own egg from those of the parasite. This is not to imply that all birds are capable of spotting a fake in the nest. Through the process of natural selection, those pairs within each host species of a parasite are favored that have a heightened sensitivity for egg recognition.

In an experiment with female African Village Weavers, ornithologists found the birds to be very adept at distinguishing between their own eggs and those of another female. Presumably, this is some sort of defense against the Diederik Cuckoo, which frequently attempts to use the weaver as its host, although it may also prevent the successful dumping of eggs by females of its own species. Only when the invading egg is a perfect match, does the weaver not cast it from the nest.

Among thirty-five species of North American birds that were studied, seven were found to recognize their own eggs and reject others. When the egg of a Brown-headed Cowbird, a nest parasite whose eggs don't necessarily resemble those of its hosts, was laid in the nests of an American Robin, Cedar Waxwing, Blue Jay, Brown Thrasher, Northern Oriole, Gray Catbird, and Eastern Kingbird, these hosts readily rejected the interloper's eggs. On the other hand, a Red-winged Blackbird, American Goldfinch, and American Mourning Dove accepted the cowbird's egg and began incubating. Assuming a parasitic egg is not rejected, the nestling parasite, like its forefathers, knows how to position itself to increase its chances of survival.

First, the eggshell of a parasite such as the Eurasian Cuckoo is 25 percent thicker and thus stronger than the shells of most other birds, enabling it to endure more abuse in the nest and still hatch. In addition, since the incubation period of the cuckoo is about twelve and a half days – compared to thirteen or fourteen days for most host species – the young parasite hatches before its nestmates. This, of course, means that the bird will be bigger and stronger than its younger foster siblings, an important advantage at feeding time.

When the cuckoo is ten hours old, an instinct – one of the most amazing in the avian world – surfaces. If an egg, a young bird, or any solid object touches a sensitive shallow depression on the scrawny cuckoo's back, the blind nestling pushes it to the rim of the nest and shoves it overboard. This unique instinct lasts for three and a half to four days, the end result being that the nest is emptied of any other eggs or young birds.

Honeyguides go a step farther. The newly hatched honeyguide emerges from its shell with a needle-sharp membranous hook on the tip of the normally pointed bill. When it comes into contact with an egg or a newly hatched foster sibling, it automatically strikes out, slashing, injuring, and eventually killing its nestmates. After about a week, the hook falls off and if a nestling is then placed in the nest with the honeyguide, it is not attacked.

The foster parent continues to care for the parasite, in some cases often longer than it would care for its own offspring. The host parent generally has such a strong urge to brood and feed young within the nest that it will neglect its own starving offspring that have been tossed overboard and concentrate solely on the ravenous young parasite. In the case of the honeyguide, the hole-nesting host simply removes its own dead young and continues to feed the honeyguide.

PREHATCHING BEHAVIOR

An incubating bird sits very still on its (and, sometimes, another bird's) eggs. Every so often the bird rises, peers at its eggs, arches its neck and draws its bill backwards over the eggs, or it may face in a different direction when it settles. These simple movements rearrange or turn the eggs in the nest, allowing those nearer the edge to be moved into the middle, so that the eggs are warmed evenly.

Some birds rotate the eggs every time they return to the nest; others do so as often as every few minutes. The African Palm Swift has an effortless method for achieving this task. The bird uses its saliva to glue its tiny nest to a palm frond, which then sways in the wind, allowing the whole egg adequate exposure when needed for cooling.

Heat distribution is not the only reason for turning the eggs. When eggs are not turned, the egg membrane may stick to the shell, interrupting the normal development of the embryo.

In the early stages of egg-laying, many birds will desert the nest readily at any sign of trouble or disturbance. Some birds such as Caspian Terns ignore an egg that has rolled from the nest, while continuing to incubate the remaining eggs. And cormorants can be such sloppy incubators that it isn't uncommon for some of these birds' eggs to roll out of the nest, falling victim to neglect or foraging gulls.

But as the clutch ages and nears hatching, most birds tend to stick as close as possible to the nest, leaving only when the situation is one of life or death. Social African barbets always leave one member of the group to incubate or perch near the nest, guarding against honeyguides, potential nest competitors, and predators.

Some intrepid birds – even when threatened – stick like glue to their eggs. The Killdeer female will hold fast to her nest, occasionally even allowing herself to be touched by a human before she will desert her brood.

The instinct to brood is a matter of survival. And, even, it would seem, stubbornness – in at least one documented case of a Bufflehead. This small, chubby duck was nesting in a rotten stump in Alberta, Canada. When a human observer intruded and tried to shoo the bird away, he had to go so far as to pry the bird off the eggs and throw her into the air before she would leave.

THE EGG

Here is what the female bird is able to accomplish in a mere twenty-four hours.

After one of the male's many sperm fertilizes the female's ripe ovum or egg, a small spot on the ovum – now a developing embryo – bursts through the ovary and slowly begins the journey down the meandering oviduct, a long and convoluted tube with elastic walls. During this journey, the egg as we know it – yolk, white, and enclosing shell – will be produced.

Peristaltic contractions of the muscle layers of the oviduct squeeze the developing egg into each of its five chambers in sequence. It is in the magnum of the oviduct that the egg receives its layer of albumen, a task taking about three hours. This albumen (the white part of the egg, as any chicken-egg eater can identify) surrounds the embryo and holds much of the water that is so important to the developing chick's survival.

When the magnum first secretes the albumen, it is a dense, jellylike layer. By the time the egg is laid, however, four separate layers of albumen are detectable. The first layer is thin and watery, a mixture within which the embryo may rotate. The next is a thicker one, surrounded by a fluid layer. The fourth is called the chalaza, a pair of dense, twisted "cords" of egg-white attached to opposite ends of the embryo and yolk. It is the job of these cords to attach the embryo to the shell and enable it to turn as the incubating bird turns the egg.

The next step in the egg-producing process occurs in the isthmus where the egg receives its two shell membranes, the first surrounding the albumen, the outer membrane firmly attached to the shell itself.

Within the uterus itself, the shell is formed over the next nineteen to twenty hours. This porous structure is made up primarily of calcium carbonate. As a rule, the thickness of the shell is proportionate to its size, with larger eggshells being thicker than smaller ones. The egg colors, if any, are produced here.

Once the shell is added, through the laying process, and thereafter over the course of its incubation, the chick must receive all its nutrients from the contents of the egg, including calcium from the shell. In many birds, the yolk, the primary source of nutrition, is almost depleted by the time the young bird hatches.

Oxygen is absorbed through the shell into the egg and carbon dioxide is expelled through blood vessels in the embryo and membrane, and out through tiny pores in the shell, for the chick must be able to breathe if it is to metabolize food.

The egg even contains a sac for the chick's waste products, allowing waste to be kept separate from the unused food sources.

When the egg is fully formed, its mother has produced a nearly perfect environment for her developing chick.

11

The Interrelationship
Between Man and Birds

In the opening pages of this book, we read of an American
Robin hatching. In ensuing chapters, we've seen how the robin
and numerous other birds live their lives. Let's say our proto-
typical robin learned its lessons in the nest and thereafter well,
flying with grace and ease, foraging in the choicest places for
the tastiest morsels. The male robin won a prime territory from
which he sang his most beautiful song. He wooed a mate and
together they raised a brood of their own. He traveled south
when the cold winds came and many times he found himself in
danger over which he triumphed.

But now, sad to say, our robin has died.

It is difficult to be sure how old this bird was. The oldest
American Robin on record thus far was a banded bird who
somehow survived eleven winters. In a perfect world, this robin
might not be such a rarity. But in the real world filled with real
hazards, any small bird such as the robin is lucky to survive, even
to venture out of the nest.

In one study of European Robins hatched in the spring,
seventy-two out of one hundred had died by August 1. Similarly,
only 22 percent of Skylarks survive the first twenty days of life
(these deaths exclude the unhatched eggs and dead embryos).
Not all birds fare so poorly; in general, the more developed the
birds at hatching, the lower the fledgling mortality. Thus, the
larger the bird when it leaves the nest, the better its chances of
survival.

As a rule, larger birds tend to live longer than smaller ones.
The oldest living Royal Albatross lived almost thirty-six years,

compared to the record-holder for Ruby-throated Humming-birds, which was five when it died. Typically, smaller birds breed during the first year and produce relatively large clutches, a way of maximizing reproduction during a short life span, whereas larger birds can afford to wait several years before breeding, and lay only a single egg, yet be capable of reproducing for decades.

Most small birds that successfully fledge are fortunate if they live to breed once, for a high proportion of young birds do not survive the first year. Most survivors breed once, or if they are lucky, again the next year. The few that live on for four, five, or more years contribute significantly to the gene pool in the next generations. In this way, natural selection greatly rewards success and severely punishes failure.

How did our robin die? That, too, is difficult to say, for the ways in which birds die are almost as limitless as human imagination, and few deaths are actually witnessed by humans. The following is just a mere sampling of the ways in which our robin might have met its fate:

- A predator pounced on it while it was preoccupied with devouring a worm.
- It starved to death.
- It succumbed to a freak spring snowstorm.
- The robin got into a dispute with another bird, was injured, and died.
- It died of disease.

The list of the possible causes of this bird's death could go on and on. And we have yet to address the part that man may have played in the robin's demise.

MAN'S ROLE IN AVIAN MORTALITY

In the name of progress we have leveled our forests, out of which have sprung subdivisions, displacing and killing vast populations of birds. Freeways crisscross the landscape and factories belch smoke into the sky and spew industrial waste into the water supply. At what cost to the birds and other wild creatures?

Man and his by-products are responsible for approximately 270 million bird deaths every year in the continental United States alone. This figure does not include perhaps an equal number of deaths indirectly caused by man's destruction of habitat.

Birds are destroyed easily, in many cases effortlessly, and often without a second thought. The great bulk of deaths occur out of sight and sound of humans, who contribute greatly to the causes of these deaths.

One way in which birds are killed deliberately is through hunting. Unlike the days when people hunted for food, today's hunter most likely does it for the sport and rarely has a real need for the food that he has killed. An estimated 5 million hunters annually stalk the marshes and open fields in search of duck, pheasant, quail, and other game birds.

Any species of bird that is unlucky enough to be among those coveted by hunters is assured a high mortality rate. In years with open hunting seasons on the Canvasback duck, the mortality rate averages 50 percent or more. But when hunting is prohibited because of a low duck population, the death rate is lowered by 14 to 21 percent. On the plus side, duck hunters pay fees and belong to organizations that help to protect the breeding grounds of the ducks, ensuring a substantial harvest.

Between 1941 and 1946, some pilots flying out of a Texas airport shot between 675 and 1,008 Golden Eagles a year. And one ornithologist witnessed the slaughter of a flock of some eight hundred Sharp-shinned Hawks flying across a road in Cape May, New Jersey, one autumn day in 1935. Hunters gathered en masse along the hawks' annual migratory route waited for the birds and then let loose with an explosion of firearms. By noon, 254 dead birds littered the pavement.

Fortunately, the public consciousness has been raised. Hunting remains popular, but it is more strictly regulated and examples of widespread slaughter such as those described would surely draw the wrath of authorities and environmentalists alike. There are special laws and regulations on the books intended to protect birds from the threat of extinction. Nonetheless, some gun-owners are careless or simply don't care what they shoot at. And all too often the targets of their fire are rarer species, even threatened ones, that are off-limits.

Even when birds are not the primary target of man's hunting efforts, they sometimes become accidental victims. Every year approximately 500,000 Thick-billed Murres are trapped in nets cast for fish, while an estimated 21,000 Redhead ducks each winter are caught in trot-lines along the Texas coast. Of this latter number, more than 3,000 drown, while 7,000 of the escapees are so badly injured that their chances of survival are low.

Cars alone kill millions of birds each year. In one study of avian road deaths in England, observers calculated that at least 2.5 million birds there were killed annually; in Denmark, a similar study estimated the number of road deaths at 3 million a year. Many drivers don't realize when they've hit a bird, and most people don't understand that bright automobile lights can blind birds flying at night.

High-tension lines, television towers, plate-glass windows, and farm machinery all take their toll.

One of the least-recognized causes of death to birds is a collision with a plate-glass window. The bird, unable to distinguish between natural flight paths and the reflective glass, slams into the window. Almost everyone reading this book probably can recall at least once when he or she was startled by the thud of a bird crashing into their picture window or against the windshield of their car. Sometimes the bird is simply stunned from the accident and is able to pick itself up off the ground and fly again. But in at least half of all such accidents, the bird quickly dies of a skull fracture and intracranial hemorrhage, and in more cases, it later succumbs to the after-effects.

It is difficult to say how often such accidents occur but the incidence is somewhere in the millions. In one study in which a single home in Illinois was monitored twenty-four hours a day for one year, sixty-one collisions were noted. If one were to extrapolate from this number, the minimum annual death rate from window collisions would be 80 million birds in the United States alone.

Other natural death traps for birds, especially nocturnal flyers, are television and radio towers, which are responsible for an estimated 1.2 million bird deaths annually. In one study, an ornithologist attempted to pick up all the birds killed by a television tower in Florida. Over a six-year period, 15,200 birds belonging to 150 species were found underneath the tower. The actual number of deaths, however, was probably much higher because many carcasses were carried off by predators before they could be counted.

Skyscrapers, with their beacon-like lights illuminating the night sky, also have proven to be killers. On the night of September 15, 1964, 497 birds were found dead on the streets and rooftops around New York's Empire State Building. As for farm machinery, it has not only destroyed many a species' natural habitat, but is responsible for the deaths of millions of adult birds and destruction of their eggs every year. In a single two-week period of mowing in the Oregon countryside, two farmers estimated they had killed between four hundred and six hundred shorebirds, ducks, and cranes plus countless numbers of nests.

Even something as seemingly benign as the noise from a supersonic aircraft can have a drastic impact on an avian population, as exemplified by the Sooty Terns on the Caribbean islands of the Dry Tortugas.

These birds had been raising between twenty and twenty-five thousand young a year. In 1969, fifty thousand pairs settled on the islands to breed but were able to rear only two hundred forty chicks. Apparently, frequent sonic booms from jet aircraft caused the birds to fly up in panic each time, which interrupted the normal rhythm of incubation and caused mass desertion of the nests.

Around John F. Kennedy Airport in New York City, thousands of Laughing Gulls that come to breed in the spring are shot by the authorities to prevent the birds from colliding with the planes.

The role of predators in avian deaths is no small one. Virtually every species has its own nemesis. For the Common Tern, it happens to be the ordinary rat. Eating the eggs of the birds as well as their young, rats have been known to wipe out an entire colony of tern chicks. At the same time, the adult terns are vulnerable to the Great Horned Owl, which in the space of one night may decapitate fifteen to twenty terns. It isn't difficult to see that in a year of an unusually large predator population, its main food source will not fare well. Over a nineteen-year period on the Dry Tortugas, a colony that began as thirty-five thousand Common Noddy Terns was virtually annihilated by rats, with only four hundred birds surviving.

Tropical birds and their eggs or offspring frequently fall victim to snakes, monkeys, weasels, and a host of other reptiles, mammals, and birds. Biting ants prey on quail nestlings. Northern Pike are major predators of young ducklings as are Snapping Turtles; and, of course, many birds prey upon other birds.

Prior to 1931, four hundred thousand birds lived on the small island of Herokapare, off the coast of New Zealand. Then cats were introduced to the island. Twelve years later only a few thousand birds remained; six species had become extinct.

Although human beings certainly have no control over most avian predators, one deadly and pervasive bird enemy under a modicum of human control is the house cat. A common sight in any suburban or small town neighborhood is a cat running across the street with its latest bird kill hanging from its mouth. And it is doubtful that we humans observe more than a small fraction of these killings since many cats hunt birds at night. Obviously, cat owners have better things to do than police their pets. But people who do own cats shouldn't go out of their way to attract birds into their yards with bird feeders, birdbaths, and stray crumbs from children's sandwiches. This only contributes to the number of birds killed by cats.

OUR POISONED ENVIRONMENT

When a hunter shoots at a duck or goose, shotgun pellets that don't find their mark end up in the water, sinking down to the bottom. In general, the hunter of average skill shoots between five and thirty twelve-gauge shells for every fowl killed. Contained within each cartridge are two hundred fifty lead pellets. This means that every acre of lake used for hunting may contain more than fifty thousand pellets of lead; in lakes with firm bottoms, the number is considerably higher. Geese and ducks commonly ingest the pellets accidentally, along with the seeds and vegetation they eat. Whether or not the lead kills the bird depends upon several factors such as the number of pellets eaten and whether they are pulverized in the gizzard. For most birds, one crushed pellet is usually enough to prove fatal.

How common is this sort of poisoning? In one study, thirty-six thousand birds representing twenty species of waterfowl were dissected after death. Scientists found lead in 6.7 percent of duck gizzards and in 1 percent of geese.

The price of an industrialized society has been high in terms of environmental health. And, unfortunately, the bad habits of the developed countries have been passed on to developing countries, where weak economies and a lack of knowledge encourage still more damage to the environment. For every creature that inhabits the earth, the risks of breathing impure air and drinking tainted water are increasing and are likely to continue to do so unless very stringent measures are taken. For many birds and other animals depending on the purity of these natural resources, time has already expired.

In Cape Town, South Africa, the air is so polluted from the use of leaded gasoline that Laughing Doves tested in one study had a mean lead content of 84.3 parts per million in their blood, compared with a lead reading of 13.1 parts per million in doves living in rural areas.

Even birds whose bodies are not obviously damaged by a polluted environment become victims as their previous habitats become unfit places in which to nest and feed. And if they are not victims directly, they may be unable to nest and rear young.

Every year coal mines – some of them abandoned – spew 3.5 million tons of acid into the streams in this country alone, causing the destruction of thousands of miles of river habitat. Because of air pollution, the chemistry of many of our country's wetlands has been so radically altered that they can no longer support the wildlife once dependent upon them. And this is in a country with considerable public awareness and some legal measures that should enable us to clean up our environment.

Anyone who has seen news reports of clean-up efforts after a major oil spill has seen at first hand what a bath in oil does to a bird. The bird becomes cold due to the oiled feathers. It tries to preen itself, which results in a good amount of petroleum being swallowed. Once inside the bird, the oil can cause gastrointestinal problems, pneumonia, liver disease, and reproductive failure. Moreover, an oiled bird may drip oil on its eggs in the nest, greatly reducing their chances of hatching. Most birds contaminated by oil are doomed to die.

On the Gulf and West coasts of the United States, the Brown Pelican – once a bird found in abundance – became almost extinct in the 1960s when its efforts to reproduce failed. Investigators found that hydrocarbon pesticide residues in the marine food chains of the coastal states interfered with the pelicans' ability to produce normal eggshells. Instead, the birds laid eggs with shells so thin that when the parent bird started to incubate them, the fragile eggs cracked under the parent's weight.

In the past fifty years, chemical advances in pesticides have done wonders for the farming industry, wiping out millions of insects and other pests that are the bane of a farmer's existence. Between 1951 and 1966, worldwide food production increased 34 percent, thanks in a considerable degree to the pesticides; in the same time period the use of pesticides increased 300 percent.

Unfortunately, many of these pesticides are not only highly toxic to insects, but remain in the environment for many years, infiltrating the soil, water, and ground vegetation.

One of the most toxic and most talked about pesticides, DDT, has now been banned in the United States though it is still being sold in some Third World countries where regulations are less strict. Here's the reason for the ban.

When a bird eats food containing DDT and the dosage is not so high so as to immediately kill it, the bird may become aggressive and extremely nervous. It may be unable to lay eggs or if it does lay, the eggs may be sterile. Some DDT-exposed birds lay eggs with shells so thin that incubating is impossible. Other birds will eat the eggs they have just laid.

Sometimes a bird will store levels of DDT in its body fat which, if contained in the nervous system, would be fatal. Some birds – if in good health – can withstand a certain amount of DDT. However, when food is scarce and the body begins living on its fat stores, the toxic pesticide is released into the bloodstream at lethal levels. Soon afterward the bird dies.

In one experiment, twenty Brown-headed Cowbirds were fed food containing DDT for thirteen days, and then untainted food for two days. Then the amount of food was reduced by 43 percent. Within four days, seven of the birds were dead. But within a control group with the same diet but without a reduction in food rations, all the birds survived.

DDT is by no means the only deadly pesticide. It is not known how many bird deaths are attributable to pesticide exposure. Even when the chemicals do not actually kill a bird, they often are so detrimental to reproduction that an entire bird population can be seriously jeopardized.

EXTINCT AND ENDANGERED SPECIES

Parrots rank among the most intelligent of birds, with mental abilities that some researchers have gauged to be on a par with those of chimpanzees and dolphins. Once believed to merely mimic human speech, parrots in recent studies have shown themselves quite capable of associative learning, seeming to understand and reply to specific questions, identify shapes and colors, name objects, and even distinguish concepts such as "bigger" and "smaller."

But unless things change, fast, many of the world's parrots could soon be little more than a memory in the future. To date, almost one-quarter of the three hundred known species of parrots are at risk of extinction.

There are several reasons why many parrots are in jeopardy. Habitat destruction is one. Typically, these birds nest in cavities found in trees in the tropics. But as more tropical forests have been destroyed by man, the birds have been displaced from their natural habitat. This, in turn, has made reproduction in many parrots species, which is precarious under the best of circumstances, even more difficult. Many parrots don't breed every year and when they do, many of the eggs and young fall victim to predators, so their populations are not growing fast enough to make up for the considerable losses.

Added to this equation of doom is the human fascination with exotic birds. The demand today for these beautiful birds is higher than ever. Trappers, often poor farmers struggling to make ends meet, comb the forests in search of the valuable birds, shipping millions of them to countries where people are willing to pay to own one. Most of the birds die in transit. In the United States alone, two hundred and fifty thousand parrots are brought in annually.

Parrots are not the only birds whose numbers have dipped dangerously low. In the past two hundred years, seventy to eighty species of birds have become extinct. Today at least one hundred species of birds are represented by fewer than two thousand individuals, placing them in great danger of extinction.

These birds include: a) the California Condor: there are about twenty in existence; b) the Whooping Crane: about forty-six are in captive breeding programs; c) the Eskimo Curlew: the numbers are uncertain, with one or two seen every few years; d) the Bachman's Warbler: a single male was last seen in 1975; e) the Ivory-billed Woodpecker: two or three were seen in 1986 and 1987, but none since then.

Birds have inhabited our planet for about 165 million years. Ornithologists have estimated that over the course of this long history up to several million species may have evolved. Today, there are about nine thousand species in the world.

A large number of species, of course, died out through the action of natural selection. Some adapted and changed, evolving into new species. Others were replaced by more adaptable species which, eventually, were replaced by other species. Still others failed to adapt and became extinct. This is the nature of the evolutionary process.

But it was not through the course of avian evolution that some 2 billion Passenger Pigeons that flew over our country in colonial times were slaughtered. Men did this, for food. Nor did nature destroy the swamp forests and extensive pine forests of the southern United States, the habitats of the Ivory-billed Woodpecker, now probably extinct there. And the Great Auk, which used to nest on islands of the North Atlantic, lives no more because its feathers and meat were once prized by humans.

In fact, all birds that have become endangered or extinct in the past two hundred years have suffered as a direct or indirect result of our activities.

Humans have destroyed habitats. In Hawaii, scientists have found the fossils of more than thirty-nine extinct birds, most of which succumbed to the extensive clearing of lowlands by the early Polynesians who came to the island, as well as to hunting. Others were eliminated from the lowlands by human-introduced diseases.

Hunters have wiped out whole populations of eagles, whose large size makes them an easy target. Some egret species were brought to the brink of extinction early in the century by hunters seeking their plumes for ladies' hats.

Colonists' introduction of animals such as cats, rats, and weasels have killed vast numbers of native species in New Zealand and around the world. Sheep, goats, deer and other mammals brought in by man have eaten foods formerly used by the birds and as a result habitats have changed. The introduction of other species of birds that compete for the same food and nesting spots has helped reduce some native species, including several of the Hawaiian honeycreepers. And other species may have come to an end because of diseases brought to them by man.

Thankfully, recent laws eliminating or restricting hunting and banning certain pesticides and other toxic materials, as well as making efforts toward the preservation of habitats, offer some encouragement for species whose numbers are dangerously low, and for many others adversely affected if not threatened outright.

For the Trumpeter Swan, aid has come in the form of suitable refuges that offer increased protection for these beautiful creatures whose numbers by 1935 had been reduced to thirty-five. Absolute control of hunting and casual shooting of the swans and public education regarding this bird also have helped. There are now more than five thousand Trumpeter Swans in existence.

In an attempt to keep the Imperial Eagle from extinction, scientists sought ways to increase reproductive success. Pairs of this eagle species usually lay a clutch of one to three eggs. When three eggs are laid, the third egg normally does not hatch or the nestling that does hatch does not survive. So ornithologists began transferring third eggs to active nests that contained only one egg. The result has been a 43 percent increase in fledgling success.

Along similar lines, captive breeding programs have been adopted by many zoos to bolster low populations of a number of rare species. Unfortunately, some birds do not breed well in captivity.

Although these efforts have had some success, much more must be done, especially in halting the destruction of natural habitats. If this is not accomplished, many more of the birds of the world will become extinct during our children's and grandchildren's lifetimes.

HUMANS AND BIRDS:
TOGETHER IN THE CHAIN OF LIFE

In the eighteenth century, the Dutch sought to maintain a monopoly of the nutmeg trade on Amboina, an island in what is now Indonesia. Their efforts were thwarted by a rather unlikely form of opposition: the fruit pigeon. These mobile pigeons moved about from island to island. The birds ate fruits from the nutmeg trees, and flew off to other islands, where they then defecated the undigested nutmeg seeds. The seeds took root and grew into trees, which effectively squelched the Dutch plans for nutmeg supremacy.

Although the Dutch were no doubt vexed by this pigeon, the story illustrates one of many ways in which birds benefit human beings.

To a considerable degree, we depend upon birds to pollinate and distribute flowering plants. In the temperate zone of North America, at least 150 species of bird-pollinated plants exist. Many birds such as hummingbirds, honeyeaters, sunbirds, orioles, and some finches fly from flower to flower, sipping or drinking nectar and seizing insects. Pollen grains bearing the male gametes of the plants adhere to the bird, which then flies off to another flower, where some pollen is unknowingly and accidentally transferred. If a bird were to flit haphazardly from one flower species to another, the chances of its transferring pollen to plants of the same species would be slight, and little fertilization would take place. But since birds tend to forage at clusters of flowers of one species at a time, the chance of pollen transfer and thus of fertilization is enhanced.

While the nectar-eaters ensure the proliferation of certain plants, pigeons, woodpeckers, thrushes, crows, and jays are among the species dispersing seeds of other plants.

An open field can be converted in time into a forest through the action of bird- and wind-dispersed fruits and seeds. Tropical seeds and fruits are more readily dispersed by birds because they are available year round.

In fact, for many years after the Dodo became extinct on the Indian Ocean island of Mauritius, no new Calvaria trees grew there. Scientists then speculated that the large seeds could only germinate after they had passed through the Dodo's gizzard. Now, Turkeys are being fed the seeds, which then can be successfully planted.

Another way in which birds indirectly help man is by eating countless billions of insects. While it is clear that the presence of birds has an overall impact on the number of insects, whether birds actually *control* insect populations is not known, because most studies have been done on farms or other modified habitats containing a reduced number of birds. Studies of pests of forest trees do show that birds, while not actually controlling insect numbers, are especially effective in preventing a build-up of certain insects. This helps to control severe outbreaks of insects that are harmful to trees.

In one study, Northern Flickers during two winters removed 64 and 82 percent, respectively, of corn-borer larvae from a corn field in Mississippi. Many birds eat insects whose larvae live on farm animals, and some such as starlings and tick-birds (African oxpeckers) actually pick ticks and other insects off the bodies of the mammals, thus helping them to avoid certain insect-borne animal diseases. Meat-eating birds such as hawks, owls, and shrikes also help control rodent populations.

Yet when the average person sees a bird, he or she probably doesn't think about how that creature is helping our greenery proliferate or about the impact it has on the number of insects and other pests. What we as human beings are more likely to see is the bird's beauty and the grace with which it travels the skies.

As our population increases and natural habitats shrink, the opportunities are rapidly waning to see the wild creatures of our world elsewhere than, for all too many of them, in zoos created by and for humans. For most of us, those birds that inhabit our backyards and city parks may represent one of our few remaining links with the natural world.

Imagine the eery silence of a forest without birdsong; or a beach without a swooping gull; or a park in which no pigeon coos.

Let us hope that these images are never realized. These fascinating and beautiful creatures have greatly enkanced the lives of countless humans. We must work to ensure that these co-beings continue to exist in all their fabulous forms, with all the habits (known and as yet unknown) that uplift us. A world without birds would indeed be a bleak place – for many of us, the equivalent of a prison.

Glossary

Altricial: Birds whose young when hatched are relatively little developed and helpless, usually blind and naked, requiring parental care for some time.

Anting: Practice in which some birds place ants on their feathers or sit on anthills and let the ants move on their bodies.

Associative learning: The way in which an individual learns by observing and imitating the actions of others.

Avian: Having to do with birds, Aves being the zoological class of birds.

Billing: The touching of bills, which in some species is part of the courtship ritual, and is a display.

Brood parasitism: The practice in which females of some species lay their eggs in other species' nests, or sometimes in the nests of other pairs of their own species.

Broodiness: A behavior pattern in which the female crouches over and spreads her ventral feathers as if to cover eggs or young, or even other objects.

Brooding: The act of sitting over nestlings or fledglings to protect them from rain or sun, and to warm or cool them.

Brood patch: An area on the bird – usually on the abdomen – that has shed its feathers and become engorged with blood capillaries in the breeding season, allowing the rapid transfer of body heat from adult to eggs or young.

Clutch: The full complement of eggs of a female bird in one nesting effort. Some birds have several clutches per year.

Dispersal: Permanent movement (emigration) of independent, usually immature or subadult individuals away from the area in which they were hatched.

Displays: Ritualized, exact behavioral postures, movements, and signals that are used as a kind of sign language in communications between members of a species and at times with other species. Displays are very diverse, and may be visual, auditory, or both.

Egg tooth: A temporary horny projection of the tip of the bill in hatching birds, usually shed rapidly.

Fledgling: A young bird that has left the nest (thus, fledged) and is still under the parents' care.

Floater: A bird of either sex in the breeding season that does not have a territory of its own, and generally wanders where it can without conflict with territorial birds.

Gizzard: The digestive organ containing grit that functions in place of teeth to grind up food. It also acts as a trap that prevents certain inedible items such as sharp bones from passing through to the more delicate parts of the intestinal tract.

Imprinting: A form of learning whereby some young birds learn to recognize and then to follow regularly a particular individual or object, a parent or parent substitute.

Incubation: The act of a bird sitting on the eggs, using its own heat to warm them so that they will develop and hatch.

Individual distance: A usually small area around an individual, different in different species, into which it will usually allow no other individual.

Lek: Sexual arena where promiscuous male birds display to attract mates.

Megapodes: Ground-living birds related to chickens, turkeys, grouse, and pheasants that use volcanically heated soil, sand, or decaying vegetation in which to lay their eggs, which thus are "incubated" without the parent sitting on them.

Migration: The two-way movement of some species between the breeding grounds and other areas used in the nonbreeding period.

Mobbing: A behavior in which individuals of one or more species join together, call at, and sometimes attack a predator to distract it and cause it to leave the general area of their breeding activities.

Modifiable behavior: Behavior that can be changed with experience, or learned.

Monogamy: A pair bond formed by one male and one female that may last one season, several seasons, or even for life.

Nestling: A term referring to hatchling birds that have yet to leave the nest; the young bird from hatchling to fledgling.

Passerine: A bird belonging to the largest order (Passeriformes), which includes more than half of all living birds and consists of relatively small-sized, mainly altricial songbirds.

Pecking order: The hierarchy of dominance in a flock, group, or cage, ranging from the most dominant bird at the top of the order to the most subordinate bird at the bottom.

Poikilothermy: A condition in which the body cannot maintain a constant temperature, and the temperature is regulated by the environment.

Polyandry: A form of polygamy in which one female mates with more than one male in a season. The male then usually builds the nest, incubates the eggs, and feeds and cares for the young.

Polygamy: The mating of one individual with more than one individual of the opposite sex.

Polygyny: A form of polygamy in which one male mates with more than one female per season.

Precocial: Birds whose young hatch well-developed and capable of moving, feeding, and other activities shortly after hatching. Precocial young hatch with a covering of down and open eyes, and are able to walk almost immediately.

Preening: The act of grooming the feathers, often using the secretion of the bird's oil gland that conditions the feathers.

Promiscuity: Breeding with several members of the opposite sex with a very short, almost nonexistent pair bond.

Releaser: A particular color, pattern, or display with a definite function in communicating with other individuals.

Social facilitation: The way in which the behavior of one animal is enhanced or modified by the actions of others.

Syrinx: Voicebox of birds, a complex structure at the base of the windpipe.

Territory: An area (small to large) that a bird defends as its own. Functions of territory include courtship, breeding, nesting, and feeding.

Unmodifiable behavior: Instinctive behavior that occurs in full form when appropriate; stereotyped.

Zugdisposition: The tendency of some migratory birds to feed beyond their usual amount, allowing large fat stores to accumulate prior to migration.

Zugstimmung: Changes in behavior and condition that are undergone before a bird can begin and sustain long migratory flight.

Zugunruhe: A twice-yearly cycle of nocturnal restlessness that occurs in migratory birds when they are physiologically stimulated and conditioned to migrate.

For Further Reading

If you would like to read more about bird behavior, the following is a list of recommended books:

Alcock, John. *Animal Behavior.* Sunderland, MA: Sinauer Associates, Inc., 1979.

Brown, Jerram. *Helping and Communal Breeding in Birds.* Princeton, NJ: Princeton University Press, 1987.

Burton, Robert. *Bird Behavior.* New York: Alfred A. Knopf, 1985.

Campbell, Bruce, et al. *A Dictionary of Birds.* Staffordshire, England: T. & A. D. Pyser Ltd., 1985.

Cemmick, David. *Kakapo Country.* London: Hodder and Stoughton, 1987.

Collins, Nicholas, and Elsie Collins. *Nest Building and Bird Behavior.* Princeton, NJ: Princeton University Press, 1984.

Ehrlich, P. R., D. S. Dobkin, and D. Wheye. *The Birder"s Handbook.* New York: Simon & Schuster.

Faaborg, John. *Ornithology: An Ecological Approach.* Englewood Cliffs, NJ: Prentice-Hall, 1988.

Forshaw, Joseph. *Encyclopedia of Birds.* New York: Smithmark Publishers, Inc., 1991.

Gill, Frank B. *Ornithology.* New York: W. H. Freeman and Co., 1990.

Goodfellow, Peter. *Birds as Builders.* New York: Arco Publishing Co., 1977.

Lack, David. *The Life of a Robin.* London: H. F. & G. Witherby Ltd., 1943.

National Geographic Society. *Field Guide to the Birds of North America.* Washington, DC: National Geographic Society, 1983.

Perrins, Christopher. *Birds.* New York: Harry N. Abrams, Inc., 1976.

Skutch, Alexander F. *Parent Birds and Their Young.* Austin, TX: University of Texas Press, 1976.

———. *Helpers at Birds' Nests.* Iowa City, University of Iowa Press, 1987.

Stacey, Peter. *Cooperative Breeding in Birds*. Cambridge, England: Cambridge University Press, 1990.

Stokes, D. W., and L. Q. Stokes. *A Guide to Bird Behavior.* Volumes I–III, Stokes Nature Guides. Boston, MA: Little, Brown and Co., 1979–1989.

Welty, Joel C., and L. Baptista. *The Life of Birds*, 4th Ed. New York: Saunders College Publishing, 1988

Wiens, J., *The Ecolgy of Bird Communities,* (2 vols.). London: Cambridge University Press, 1989.

INDEX

Note: Page numbers in italics refer to illustrations.

ABOUT THE AUTHOR

Lester L. Short is Lamont Curator of Birds at the American Museum of Natural History. Educated as a zoologist, Dr. Short earned his doctorate at Cornell University. He was chief of the bird section of the bird and mammal laboratory at Smithsonian Institution in Washington, D.C., before coming to the Museum in 1966.

Dr. Short is the author of *Woodpeckers of the World* and of more than two hundred and fifty scientific and scholarly articles and reports, mainly based upon his bird research on six continents, many in conjunction with Jennifer Horne, his Kenya-based ornithologist spouse.

ABOUT THE MUSEUM

One of the world's great treasures, the 124-year-old American Museum of Natural History has been referred to as a library of life on the planet. It is the world's largest privately operated science museum and attracts some 3 million visitors annually drawn from all over the world.

Among its incomparable collections are more than 36 million artifacts and specimens, which include more dinosaurs, birds, spiders, mammals, fossils, and whale skeletons than any other museum. Equally important, though less well known to the general public, the Museum's work in scientific research involves more than 200 scientists and researchers. The American Museum of Natural History plays a significant role as a gatherer, evaluator, and disseminator of information on ecological systems, to scientists and the lay public alike.